Practical Guide to Interpretive Near-Infrared Spectroscopy

Jerry Workman, Jr.
Lois Weyer

CRC Press
Taylor & Francis Group
Boca Raton   London   New York

CRC Press is an imprint of the
Taylor & Francis Group, an **informa** business

Cover graphic (arbitrary absorbance scaling). This graphical representation of a near-infrared spectrum indicates the 8 (eight) locations (shown using small blue arrows) representing C-H stretching information for 2, 2, 4-trimethyl pentane. The spectrum extends from about 690 nm to 3000 nm (14,493 cm-1 to 3333 cm-1). When viewing the near-infrared spectrum from **right to left** one observes the fundamental (v) combination region where a significant quantity (high rank) of molecular vibrational information is found; then the first overtone (2 v) C-H stretching region (section 2); the first overtone (2 v) combination region (section 3); the second overtone (3 v) C-H stretching region (section 4); the second overtone (3 v) C-H combination bands (section 5); the third overtone (4 v) C-H stretching region (section 6); the third overtone (4 v) combination bands (section 7); and finally the fourth overtone (5 v) C-H stretching region (section 8).

CRC Press
Taylor & Francis Group
6000 Broken Sound Parkway NW, Suite 300
Boca Raton, FL 33487-2742

© 2008 by Taylor & Francis Group, LLC
CRC Press is an imprint of Taylor & Francis Group, an Informa business

No claim to original U.S. Government works
Printed in the United States of America on acid-free paper
10 9 8 7 6 5 4 3 2 1

International Standard Book Number-13: 978-1-57444-784-2 (Hardcover)

Library of Congress Cataloging-in-Publication Data

Workman, Jerry.
    Practical guide to interpretive near-infrared spectroscopy / Jerry Workman and Lois Weyer.
        p. cm.
    Includes bibliographical references and index.
    ISBN 1-57444-784-X (alk. paper)
    1. Near infrared spectroscopy. I. Weyer, Lois. II. Title.

QD96.I5W67 2007
543'.57--dc22   2007010011

**Visit the Taylor & Francis Web site at
http://www.taylorandfrancis.com**

**and the CRC Press Web site at
http://www.crcpress.com**

Dedication

———————

To the ones that have gone before us and upon
whose work our work is based
To the ONE who so marvelously engineered this
universe of mystery, symmetry, and magnificent order.

Jerry Workman

To the supportive and inspiring teachers and professors, including Tom
Medwick, Phyllis Dunbar, Barbara Irwin, Steve Brown.
To Steve Prescott, who let me take background materials home.
To Dorothy Tunstall for past help, and Linda Perovich for current
assistance.
To encouraging friends including Howard Barth and many in the NIR
community.
To Dorothy Delker, Caroline Herschel, and the rest of the many female
scientists of the past and present.
To the women in the Hercules Library, past and present.
To my family members, especially Gordon, Geoff, and Nathan.

Lois Weyer

Table of Contents

Preface

Qualitative and quantitative near-infrared (NIR) spectroscopic methods require the application of multivariate calibration algorithms commonly referred to as chemometric methods to model spectral response to chemical or physical properties of a calibration, teaching, or learning sample set. The identification of unique wavelength regions where changes in the response of the near-infrared spectrometer are proportional to changes in the concentration of chemical components, or changes in the physical characteristics of samples under analysis, is required for a scientific understanding of cause and effect, even for routine method development.

The first step to developing an analytical method using NIR is to measure a spectrum of the sample using an NIR spectrophotometer. It is helpful to note that the near-infrared spectrum obtained by using a spectrophotometer is the result of the convolution of the measuring instrument function with the unique optical and chemical characteristics of the sample measured. The sample participates as an optical element in the spectrometer. The resultant spectrum contains information specific to the molecular vibrational aspects of the sample, its physical properties, and its unique interaction with the measuring instrument. Relating the spectra to the chemical structure of the measured samples is referred to as spectra–structure correlation. This correlation or interpretation of spectra converts the abstract absorption data (spectrum) into structural information representing the molecular details about a measured sample. Interpretive spectroscopy of this sort provides a basis for the establishment of known cause-and-effect relationships between the spectrometer response (spectrum) and the molecular properties of the sample.

The exclusive use of chemometrics alone provides a weak basis for analytical science. When performing multivariate calibrations, analytically valid calibration models require a relationship between X (the instrument response data or spectral data) and Y (the reference data); probability tells us only if X and Y "appear" to be related. If no cause–effect relationship exists between X and Y, the analytical method will have no true predictive significance. Interpretation of NIR spectra provide the knowledge basis for understanding the cause-and-effect of molecular structure as it relates to specific types of absorptions in the NIR. Interpretive spectroscopy is a key intellectual process in approaching NIR measurements if one is to achieve an analytical understanding of these measurements. This book represents our best effort to provide the tools necessary for the analyst to interpret NIR spectra.

Although there have been no books written specifically on the interpretation of NIR spectra, a number of classic reviews have been published. A few of these have included charts or spectra showing specific band absorptions.

The most important early reviews covered the 1920s through the 1950s, a time when new commercial UV-Vis-NIR instruments became available and interest in the spectral region increased in both academia and industry. These included reviews by Wilbur Kaye[1] of Tennessee Eastman, Owen Wheeler[2] at the University of Puerto Rico, and Robert Goddu[3] of Hercules Powder Company (now Hercules Incorporated). In addition, Wheeler cited an earlier review by Joseph W. Ellis[4] summarizing NIR papers published before 1929, and Lauer and Rosenbaum[5] reviewed band assignments in 1952.

Wheeler's article includes a table of approximate theoretical wavelengths of overtones, such as the C–H stretch first overtone at 1.7 microns, the second overtone at 1.1 microns, the third at 0.85, and the fourth at 0.7. Kaye's 1954 article provided a chart of the spectra-structure correlations and approximate absorptivities that were available at that time. The same article also provided very detailed band assignment tables and spectra for a few specific compounds. In 1960, Goddu and Delker[6] published a more complete chart that has been widely reproduced. It was based on literature references and personal knowledge from a series of studies and publications.

During the 1960s and 1970s, NIR was not studied as extensively, although another important review was written by Kermit Whetsel[7] in 1968 and some work did continue at specific institutions. A chapter on NIR interpretation was included in Robert Conley's text on infrared spectroscopy published in 1966.[8] This was the time period in which Karl Norris at the USDA began his research that would lead to the greatly renewed interest in this spectral field, but there was initially a disconnect between the new agricultural applications of NIR and the earlier academic and chemical industry interests.

There were some specialized reviews during this time period. For example, reviews on hydrocarbon NIR spectra were written by Bernhard and Berthold,[9] and Tosi and Pinto.[10] An atlas of 1000 NIR spectra of a wide selection of general chemicals was published in 1981.[11] Although the spectral bands are not interpreted, the atlas provides an excellent reference collection.

As the new rapid scan, optically optimized instruments became popular in the food and agricultural industries and began to be noticed by other industries, several reviews were written to bridge the knowledge gap. These include those by Weyer,[12] Honig,[13] and Stark, Luchter, and Margoshes.[14] An important book, first published in 1987 by Norris and Williams,[15] has a chapter on chemical principles, including band assignments. There is also interpretive information in Osborne and Fearn's[16] food analysis book. More recent extensive reviews on spectral interpretation include those by Workman[17] and Weyer and Lo.[18]

The shortwave near-infrared spectral region was specifically discussed by Schrieve et al.[19] They referred to synonyms such as "the far-visible," the "near, near-infrared," or the "Herschel infrared" to describe the range of approximately 700 to 1100 nm of the EMS (electromagnetic spectrum). The authors cite the increased interest of this spectral region to spectroscopists, particularly those involved with implementing process near-infrared measurements. More details regarding this spectral region are also described in a comprehensive handbook by Workman.[20]

Jerry Workman
Madison, Wisconsin

Lois Weyer
Landenberg, Pennsylvania

REFERENCES

1. Kaye, W., Near-infrared spectroscopy: a review, spectral identification and analytical applications, *Spectrochim. Acta*, 6, 257–287, 1954.
2. Wheeler, O.H., Near-infrared spectra of organic compounds, *Chem. Rev.*, 629–666, 1959.
3. Goddu, R.F., Near-infrared spectrophotometry, in *Advances in Analytical Chemistry and Instrumentation*, Vol. 1, Reilly, C.N. (Ed.), Interscience Publishers, New York, 1960, pp. 347–424.
4. Ellis, J.W., Molecular absorption spectra of liquids below 3 microns, *Trans. Faraday Soc.*, 25, 888–898, 1929.
5. Lauer, J.L. and Rosenbaum, E.J., Near-infrared absorption spectrometry, *Appl. Spectrosc.*, 6(5), 29–46, 1952.
6. Goddu, R.F. and Delker, D.A., Spectra–structure correlations for the near-infrared region, *Anal. Chem.*, 32, 140–141, 1960.
7. Whetsel, K.B., Near-infrared spectrophotometry, *Appl. Spectrosc. Rev.*, 2(1), 1–68, 1968.
8. Conley, R.T., The near-infrared region, in *Infrared Spectroscopy*, Allyn and Bacon, Boston, MA, 1966, pp. 220–239, chap. 7.
9. Bernhard, B. and Berthold, P.H., Application of NIR spectrometry for the structural group analysis of hydrocarbon mixtures, *Jena Rev.*, 20(5), 248–251, 1975.

10. Tosi, C. and Pinto, A., Near-infrared spectroscopy of hydrocarbon functional groups, *Spectrochim. Acta*, 28A, 585–597, 1972.

11. Hirschfeld, T. and Hed, A.Z., *The Atlas of Near-Infrared Spectra*, Sadtler Research Laboratories, a Division of Bio-Rad Laboratories, Philadelphia, PA, 1981.

12. Weyer, L.G., Near-infrared spectroscopy of organic substances, *Appl. Spectrosc. Rev.*, 21(1–2), 1–43, 1985.

13. Honig, D.E., Near-infrared analysis, *Anal. Instrum.*, 14(1), 1–62, 1985.

14. Stark, E., Luchter, K., and Margoshes, M., Near-infrared analysis (NIRA): a technology for quantitative and qualitative analysis, *Appl. Spectrosc. Rev.*, 22(4), 335–399, 1986.

15. Murray, I. and Williams, P.C., Chemical principles of near-infrared technology, in *Near-Infrared Technology in the Agricultural and Food Industries*, Williams, P. and Norris, K. (Eds.), American Association of Cereal Chemists, St. Paul, MN, 1987, pp. 17–31.

16. Osborne, B.G. and Fearn, T., *Near-Infrared Spectroscopy in Food Analysis*, Longman Scientific and Technical, Harlow, 1986, pp. 29–41.

17. Workman, J.J., Interpretive spectroscopy for near-infrared, *Appl. Spectrosc. Rev.*, 31(3), 252–320, 1996.

18. Weyer, L.G. and Lo, S.-C., Spectra–structure correlations in the near-infrared, in *Handbook of Vibrational Spectroscopy*, Vol. 3, Chalmers, J.M. and Griffiths, P.R. (Eds.), John Wiley and Sons, Chichester, 2002, pp. 1817–1837.

19. Schrieve, G.D., Melish, G.G., and Ullman, A.H., The Herschel-infrared — a useful part of the spectrum, *Appl. Spectrosc.*, 45, 711–714, 1991.

20. Workman, J., Short-wave near-infrared spectroscopy, in *Handbook of Organic Compounds*, Vol. 1, Academic Press, 2001, pp. 133–141.

1 Introduction to Near-Infrared Spectra

1.1 MOLECULAR SPECTRA

Molecular spectra result from the periodic motions (or vibrational modes) of atomic nuclei within their respective molecules. These nuclei move together or apart along a straight-line vector; and they rotate; they vibrate; they wag; and they bend relative to their centers of gravity. The vibration and bending of molecules exhibit vibrational spectroscopic activity that may be measured using any number of spectroscopic techniques, including the near-infrared, mid-infrared, far-infrared (terahertz), and Raman spectroscopy. The resultant spectra from these molecular vibrational measurement techniques are highly structured and complex. The process of understanding or characterizing this complexity into the spectra–structure correlations for near-infrared spectra is the purpose of this book.

The energy level in a molecule is described as the sum of the atomic and molecular motions due to translational, rotational, vibrational, and electronic energies. Translational energy has no effect on molecular spectra, whereas the other motions do affect the spectral characteristics. Rotational energy is proportional to the angular velocity of rotation for each molecule. Electronic energy in molecules and their various quantum numbers are described via the Pauli principle and are beyond the scope of this work. We will restrict our discussion to the vibrational energy levels and use the application of what is learned in this model as a basis for our specific structure-correlation characterization of near-infrared (NIR) spectra.

1.2 VIBRATIONAL ENERGY LEVELS

Note that a molecule with N atoms has three degrees of freedom for motion (3N). For all three-dimensional objects, there are three axes of translation (x, y, z) and three axes of rotation (as there are three axes of inertia). If we eliminate these six kinds or types of motion (as they are nonvibrational), we are left with 3N-6 vibrational types of motion. (Note: If the molecule is linear like many polymers, there are 3N-5 types of motion.) Each other kind of motion is vibrational in nature and has a specific frequency associated with it. As long as the bonds do not break and the vibrations have motions (amplitudes) of about 10–15% of the average distance between atoms, the vibrations are considered harmonic. Any harmonic is considered to be the superposition of two or more vibrations of the molecule and carries the term normal vibration.

The frequency of any vibration is not dependent on amplitude. The displacement functions for these normal vibrations are described exactly as a sine-wave function. There are 3N-6 normal vibrations in a molecule, and their respective vibrational frequencies are called the fundamental frequencies of the molecule. Symmetry of the molecule is the single factor most important in determining both frequency and amplitude of a molecular vibration. The selection rules discussed later will place a heavy emphasis on symmetry.

1.3 SPECTRAL RESPONSE AND MOLECULAR CONCENTRATION

Harmonic vibrations follow a functional description such that the type of vibration determines the frequency at which it absorbs NIR energy. The amplitude of the absorption at any particular wavelength or wavenumber is determined by its absorptivity and the number of molecules encountered within the beam path of the measuring instrument. It is assumed that a change in spectral response is related to a concentration as described by the Bouguer, Lambert, and Beer relationship, most often termed Beer's law. The Beer's law relationship is described as: the absorbance (A, AU, or signal strength) of an analyte being measured using a spectrophotometer is equivalent to the product of the absorptivity (ε) of a specific type of molecular vibration; the concentration (c) of the molecules in the measurement beam; and the pathlength (l) of the sample holder within the measurement beam. This relationship between measured spectral signal and concentration of a molecule is most often expressed as

$$A = \varepsilon c l \tag{1.1}$$

where ε is the molar absorptivity (referred to as molar extinction coefficient by earlier physicists) in units of liter $\cdot$ mole^{-1} $\cdot$ cm^{-1}; c is the concentration of molecules in the spectrometer beam in units of mole$\cdot$ liter^{-1}; and pathlength is the thickness in units of centimeters of the measured sample at a specific concentration. The absorptivity for any specific molecule type is calculated by careful measurements of the absorbance of a compound generally diluted in a suitable organic solvent, and by applying the relationship

$$\varepsilon = \frac{A}{cl} \tag{1.2}$$

Note that for transmittance (where T = 0.0–1.0) and percent transmittance (where T = 0–100.0) spectroscopy, a more complete delineation of the relationships between the various terms is contained in an expression such as

$$T = \frac{I}{I_0} = 10^{-\varepsilon cl} \Rightarrow Abs. = A = -\log_{10}\left(\frac{I}{I_0}\right) = -\log_{10} T = \varepsilon c l \tag{1.3}$$

Here the symbols I and I_0 represent the attenuated energy detected after sample interaction and the initial energy incident to the sample, respectively. For reflectance (where R = 0.0 to 1.0) and percent reflectance (where R = 0.0 to 100.0) spectroscopy, the various relationships are expressed as

$$R = \frac{I}{I_0} = 10^{-\varepsilon cl} \Rightarrow Abs. = A = -\log_{10}\left(\frac{I}{I_0}\right) = -\log_{10} R = \varepsilon c l \tag{1.4}$$

The relationship where the change in intensity (I) of the transmitted or reflected light from a sample is a function of the change in pathlength (l) of the sample as expressed by the absorptivity (ε) of a specific analyte (or molecular substance) and its concentration (c) is denoted by

$$-\frac{\partial I}{\partial l} = \varepsilon c I \tag{1.5}$$

Modern spectrophotometers utilize these assumptions for making spectroscopic measurement and generally display spectroscopic data as transmittance (T), reflectance (R), and absorbance A (y-axis or ordinate axis), versus wavelength (nanometers, microns) or wavenumber (cm^{-1}) (as x-axis, or abscissa axis).

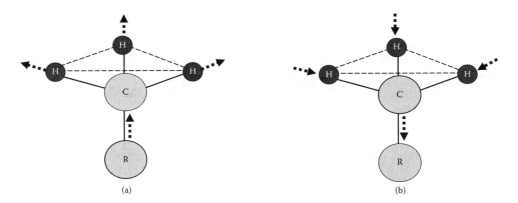

FIGURE 1.1 Methyl symmetrical stretching of $CH_3 \rightarrow v_s$. All atoms are moving toward a similar external direction.

1.4 NOMENCLATURE OF MOLECULAR VIBRATIONS

There are multiple types of molecular vibrations that absorb at unique wavelengths or frequencies of near-infrared energy depending upon the bond type. Several normal (or normal mode) types of molecular vibrations active within the NIR region are illustrated in the following figures. Each of these types of vibrations has a unique frequency where absorption occurs. The location of these frequencies and the associated molecular structures (spectra–structure correlations) are the purpose of this book.

1.5 STRETCHING VIBRATIONS

Common nomenclature used in near-infrared spectroscopy indicates that such molecular vibration is a variation of the *length* of the bond and is indicated by the symbol v. If the group consists of three atoms, instead of two, it can have two types of stretching: asymmetric and symmetric. In order to distinguish the two types of stretching, a subscript is introduced: "s" (for symmetrical stretching) and "a" (for asymmetrical stretching). Figures 1.1 through 1.6 demonstrate stretching vibrational motions in C−H-containing molecules.

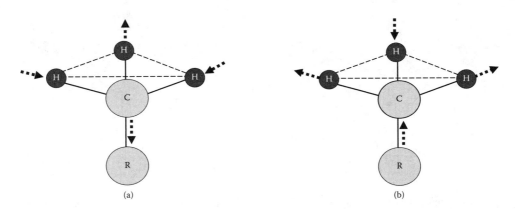

FIGURE 1.2 Methyl asymmetric stretching of $CH_3 \rightarrow v_a$. Atoms are moving in opposite directions with respect to the carbon atom.

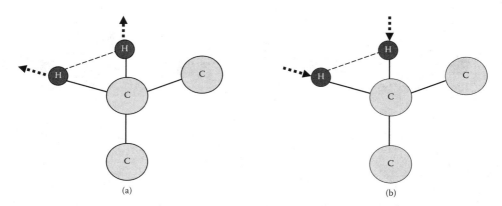

FIGURE 1.3 Methylene symmetrical stretching of $CH_2 \rightarrow \nu_s$. (In this movement the two hydrogen atoms move away from the carbon atom at the same time in the outward movement and approach each other at the same time in the return movement.)

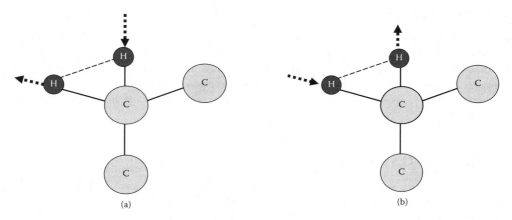

FIGURE 1.4 Methylene asymmetric stretching of $CH_2 \rightarrow \nu_a$. (In this movement, while one hydrogen is moving away from the carbon atom, the other is approaching it.)

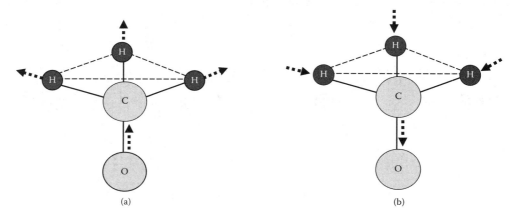

FIGURE 1.5 Symmetrical stretching of $O-C-H \rightarrow \nu_s$. (In this movement, the two hydrogen atoms move away from the carbon atom at the same time in the outward movement and approach it at the same time in the return movement.) The oxygen atom is moving in a likewise external direction as the hydrogen atoms with respect to the carbon atom.

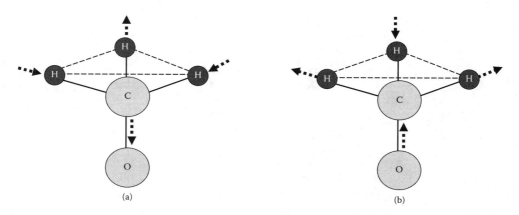

FIGURE 1.6 Asymmetric stretching of O−C−H → v_a. (In this movement, while one hydrogen atom moves away from the carbon atom, the other two approach it.) In the second type of asymmetrical motion, two hydrogen atoms move away from the carbon atom while the other moves toward the carbon atom.

1.6 BENDING

This is a variation or change in the angle of a bond occurring on the plane of the bond as indicated with the symbol δ. Figures 1.7 through 1.10 show bending molecular motions.

If the molecular group consists of four atoms (e.g., one carbon and three hydrogen atoms), instead of three (e.g., one carbon and two hydrogen atoms), they can have two types of bending: symmetrical and asymmetrical.

Symmetrical bending is indicated by adding the subscript s, and asymmetric the subscript a.

In symmetrical bending (Figure 1.10), the angle 4–1–3, the angle 4–1–2, and the angle 3–1–2 are increased at the same time in the outward movement, and are tightened at the same time in the return movement. This is a movement that simulates the opening and closing of an umbrella; it is also called *umbrella* bending.

In asymmetric bending, one of the three angles is tightened (in the outward movement), and the other two are increased; and *vice versa* in the return movement.[1]

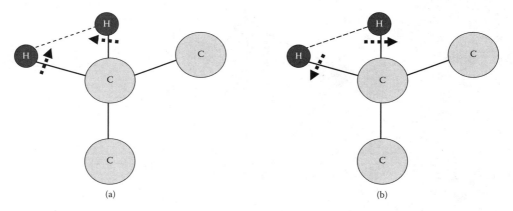

FIGURE 1.7 Methylene bending of CH_2 → δ. The two methylene hydrogen atoms move toward each other or away from each other simultaneously.

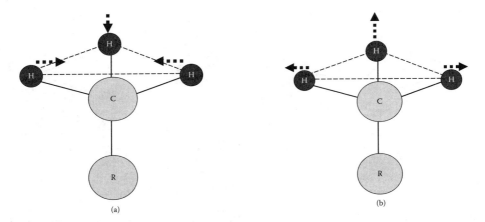

FIGURE 1.8 Methyl symmetrical bending of $CH_3 \rightarrow \delta_s$.

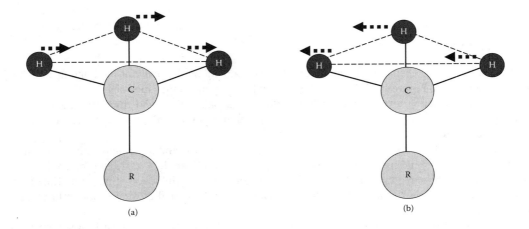

FIGURE 1.9 Methyl asymmetric bending of $CH \rightarrow \delta_a$.

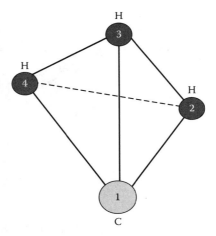

FIGURE 1.10 Internal umbrella (symmetrical) bending motion of tetrahedron CH_3.

1.7 BEGINNING GROUP THEORY FOR NEAR-INFRARED SPECTRA

Group theory is used to solve problems in chemistry specifically related to the symmetry properties of a molecule. Near-infrared absorption bands are described using the vibrational energy of the molecules measured. Molecular motion relative to the atomic nuclei is described using the x, y, z (Cartesian) coordinate system; for each atom there are three degrees of freedom for the motions allowed. If we have N atoms, then there are three times N degrees of freedom. A more useful way to describe the coordinate space of molecules is by the use of group theory. Group theory requires some basic mathematics and an understanding of the modern theories of the three-dimensional shape of molecules. Symmetry can be surmised to have an effect on the energy levels of a molecule simply by examining its spectrum. This applies to infrared, Raman, far-infrared (terahertz), near-infrared, and ^{1}H-NMR spectra.

It is not very satisfying nor quantitative to argue that the hydrogen atoms in a symmetrical molecule are all in the same "chemical" or atomic/molecular environment and thus all should "look" the same spectrally. This is certainly true but neither precise nor informative. The rules of symmetry can be precisely and mathematically described, and this is the object of group theory in its application for prediction of spectroscopic complexity based on energy levels. The first concepts and terms that will help to introduce group theory include rotational axes, reflection planes, inversion centers, improper rotational axes, and point groups. The symmetry of a molecule can be described in mathematical notation in terms of its symmetry operations. Excellent references describing group theory include Reference 2 and Reference 3.

1.8 ROTATIONAL AXES

Operations of rotation about an axis, represented by a central atom in a molecule, are an important concept in group theory. A completely symmetrical molecule can be rotated 180 degrees about its central axis and produce a molecule that is "indistinguishable" from its original; a 360-degree rotation produces an "identical" molecule. Note that this indicates that labeled molecules would be in different positions relative to a fixed observer for the 180-degree rotation, and be in identical positions for two successive 180-degree rotations. The order of these rotational axes are represented mathematically by the ratio of 360/r, where r is the number of degrees of rotation around an axis that produces an indistinguishable molecule. For example, an Sp3 molecule requires a 120-degree rotation along the long axis of the molecule and is thus 3rd-order; a molecule such as cyclohexane or benzene, when rotated around its principal axis (the axis perpendicular to the plane of the molecule), would require a 60-degree rotation and therefore has an order of 6.

1.9 REFLECTION PLANES

The concept of reflection planes refers to symmetry assessed using the mirror-image approach: an imaginary double-sided mirror is used to bisect the molecule along various planes, and the reflection from each side of the mirror is used to reconstruct the other side of the molecule. Thus, once this mental reconstruction or visualization of the molecule is completed, a symmetrical molecule will be indistinguishable from the original when using this method.

1.10 INVERSION CENTERS

Inversion centers, referred to as *i,* are another instrument for demonstrating the symmetry of a molecule. Unlike the Rotational and Reflection methods for which a molecule may have many axes or planes, there is only one inversion center per molecule. For this exercise to test symmetry, imagine that each molecule has a center of mass with bonds at opposite ends of the molecule. If the atoms at the ends of these bonds are inverted or switched with one another, does the inverted form of the molecule remain indistinguishable as compared to the original? If so, the molecule is symmetrical.

1.11 IMPROPER ROTATIONAL AXES

This is a symmetry operation needed on a molecule where one of the atoms lies along the rotational axis and cannot be transformed into the position of other atoms on the molecule simply by rotation or reflection methods. In this case, the improper rotation axis must be used to bring this atom into the position of other atoms so it can be transformed (by rotation or reflection) into the position of the other atoms. Methane can be used as an example molecule.

1.12 POINT GROUPS

The point group system is a formal nomenclature method for designating symmetry operations. *Point* indicates the single point in space that each symmetry operation passes through; this can be either a central atom within a molecule or a center of mass for a molecule without a central atom. The term *group* indicates a group of symmetry operations around the designated point. There is a set of symmetry rules that are used to classify a molecule by point groups.

1.13 DEFINING A GROUP USING MATRIX OPERATIONS

A group for this discussion may be described as a formation of atoms interacting as a molecular association and having a common set of rules whereby motion is allowed under specific prescribed conditions. There are verbal descriptive definitions for such groups, but mathematics provides a precise description. The mathematics used include the use of basic matrix algebra. Some definitions are useful: an *identity* operation is the result of a combination of events; a combination of events follows *associative properties* (in a mathematical sense); and there is an *inverse* operation for all events. So, to give mathematical definitions for these terms using matrix notation is extremely useful. Matrix multiplication is also used because of its importance in describing group interactions.

All molecules have symmetry elements in their structures. These structures have only certain potential motions allowed around any particular plane or rotational axis. For example, with crystal structures, there is a limit of 32 possible combinations of symmetry elements (or motions) allowed. These individual allowed motions are also termed *point groups*.[2] There are 32 groups, plus some additional groups that have specific axes of symmetry, for most molecules. Each of the groups has specific names and prescribed motions. These possible rotational motions are represented or expressed in matrix notation and described by using the standard matrix operations shown in the following text. The reader is referred to Chapter 3 and Chapter 14 of Reference 2 (Colthup, Daly, and Wiberley's classic infrared text) for a thorough explanation. A separate source providing a working tutorial on the subject is found in Reference 3 (Walton). Walton summarizes group theory with eight points, noting that symmetry elements are comprised of rotational axes, reflection planes, inversion centers, and improper rotation axes as illustrated by the characters in Figure 1.11. More details and descriptions of these molecular properties are to be found in Reference 2 and Reference 3.

1.14 PRACTICAL ASPECTS FOR SPECTRAL MEASUREMENTS

A variety of sample presentation methods are available to the analytical scientist. These include transmittance (straight and diffuse), reflectance (specular and diffuse), transflection (reflection and transmittance), and interactance (a combination of reflectance and transmittance). Pathlength selection

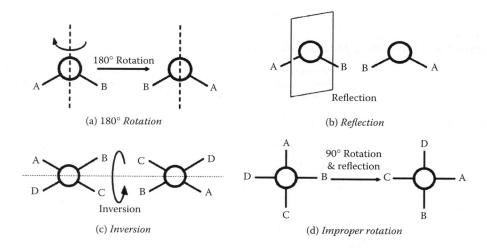

(a) *180° Rotation*

(b) *Reflection*

(c) *Inversion*

(d) *Improper rotation*

FIGURE 1.11 The symmetry operations of a molecule.

for optimum near-infrared measurements involves the following: for the shortwave near-infrared (SW-NIR) region of 12,500 to 9091 cm^{-1} (800–1100 nm), pathlengths ranging from 5–10 cm are typically used. For longwave near-infrared (LW-NIR) or 9091 to 4000 cm^{-1} (1100–2500 nm), common pathlengths for hydrocarbons include 0.1–2 cm or 1–20 mm.

The sample presentation geometries most often used in near-infrared measurements are given in Table 1.1. For this table, the abbreviations used designate transmittance (T), diffuse transmittance (DT), and diffuse reflectance (DR). Transmittance and diffuse transmittance are the identical sample presentation geometry, but the sample for DT is a series of particles allowing light to penetrate without significant backscatter. DR is used for samples of infinite optical thickness that will not transmit light if T or DT is used. The ideal surface for DR is a Lambertian or isotropic surface where light is reflected diffusely or evenly throughout a hemisphere above the surface.

Samples exhibit different optical characteristics that must be considered in order to optimize spectroscopic measurement. Standard clear samples or solutions are measured using transmission spectroscopy. Highly colored samples are generally measured using transmission spectroscopy unless the optical density exceeds the linear range of the measuring instrument. At this point, either dilution or reducing the pathlength is preferred.

Fine scattering particulates of 5 to 25 times the measuring wavelength are measured using diffuse transmittance or diffuse reflectance methods. The scattering produced by some of the

TABLE 1.1
Sample Presentation Methods for Polymers and Rubbers

Sample Type	Sample Presentation	Comments
Liquids (lower viscosity)	T and DT	Maintain Instrument within linear measurement range, that is, generally less than 1.5 AU
Slurries (high viscosity)	T and DT	Maintain Instrument within linear measurement range, that is, generally less than 1.5 AU
Solids and smaller particles	DR	For DR use infinite optical thickness
Webs	DR and DT	For DR use infinite optical thickness
Pellets (large particles)	DR and DT	For DR use infinite optical thickness

TABLE 1.2
Relative Band Intensities: MIR vs. NIR for C–H Stretch

Band	Wavenumber (Wavelength) Region	Relative Intensity	Recommended Sample Cell Pathlength (for Liquid Hydrocarbons)
Fundamental (v)	2959–2849 cm⁻¹ (3380–3510 nm)	1	0.1–4 mm
First overtone (2v)	5917–5698 cm⁻¹ (1690–1755 nm)	0.01	0.1–2 cm
Second overtone (3v)	8873–8547 cm⁻¹ (1127–1170 nm)	0.001	0.5–5 cm
Third overtone (4v)	11,834–11,390 cm⁻¹ (845–878 nm)	0.0001	5–10 cm
Fourth overtone (5v)	14,493–12,987 cm⁻¹ (690–770 nm)	0.00005	10–20 cm

reflected light creates a pseudo-pathlength effect. This effect is compensated for by using scatter correction data-processing methods for quantitative measurements, sieving the particles, or grinding to improve particle size uniformity.

Large scattering particulates on the order of 100 times the measuring wavelength in diameter present a challenge for measurements as the particles intercept the optical path at random intervals. Signal averaging can be employed to compensate for random signal fluctuations. Reflection spectroscopy can be used to measure the size, velocity, and concentration of scattering particulates within a flowing stream.

Very high absorptivity (optically dense) materials with absorbances above 4–6 AU are difficult to measure accurately without the use of high-precision photometric instruments, such as a double monochromator system having stray light specifications below 0.0001% T. Measurements can be made with extremely slow scanning speeds and by opening the slits during measurement. These measurements should be avoided by the novice unless high-performance instrumentation and technical support are available. Table 1.2 shows the relative absorption-band intensities for each particular band as well as general spectral windows for fundamental through fourth overtone C–H stretching regions.

1.15 TYPES OF NEAR-INFRARED ABSORPTION BANDS

Infrared energy is the electromagnetic energy of molecular vibration. The energy band is defined for convenience as the near-infrared covering 12,821 to 4000 cm⁻¹ (780–2500 nanometers); the infrared (or mid-infrared) as 4000 to 400 cm⁻¹ (2500–25,000 nm; and the far-infrared (or terahertz) from 400 to 10 cm⁻¹ (25,000–1,000,000 nm). Table 1.3 illustrates the region of the EMR (electromagnetic radiation) spectrum referred to as the NIR region. The table shows the molecular interactions associated with the energy frequencies (or corresponding wavelengths) of the various regions.

Specific molecular bonds most active in the NIR are listed here, with X–H bonds being the more active and intense.

C=O from aldehydes
C=O from amides
C=O from carboxylic acids
C=O from esters

TABLE 1.3
Spectroscopic Regions of Interest for Chemical Analysis

Region	Wavenumbers/ (Wavelength)	Characteristic Measured
Ultraviolet	52,632–27,778 cm⁻¹ (190–360 nm)	Electronic transitions: delocalized Pi electrons
Visible	27,778–12,821 cm⁻¹ (360–780 nm)	Electronic transitions: color measurements
Near-Infrared (NIR)	14,493–3333 cm⁻¹ (690–3000 nm) or 12,821–4000 cm⁻¹ (780–2500 nm)ᵃ	Overtone and combination bands of fundamental molecular vibrations, especially stretching and bending (some deformation as well)
Infrared (IR)	4000–400 cm⁻¹ (2500–25,000 nm)	Fundamental molecular vibrations: stretching, bending, wagging, scissoring
Far-Infrared (FIR or Terahertz)	400–10 cm⁻¹ (2.5×10^4 to 10^6 nm)	Molecular rotation

ᵃ Official ASTM International Definition.

C=O from ketones
C–H from aldehydes
C–H from alkanes
C–H from alkenes
C–H from alkynes
C–H from aromatic compounds
C–N from amines, alkyl
C–N from amines, aromatic
C–O from alcohols, ethers, and esters
N–H from amides
N–H from amines
NO_2 from nitro groups
O–H from alcohols (no hydrogen bonding)
O–H from alcohols (with hydrogen bonding)
O–H from carboxylic acids

1.16 PROPERTIES OF INFRARED–NEAR-INFRARED ENERGY

Light has both particle and wave properties; quantum theory tells us that the energy of a light particle or photon E_p is given by

$$E_p = h\nu \tag{1.6}$$

where h = Planck's constant (or 6.6256×10^{-27} erg-sec), and ν is the frequency of light (or the number of vibrations per second or in units of sec⁻¹). Thus, the energy for any specific photon can be quantified, and it is this energy that interacts with the vibrating bonds within *near-infrared active molecules*. The subsequent values for wavelength, wavenumber, and frequency for both the visible and the extended near-infrared regions are shown in Table 1.4.

TABLE 1.4
Equivalent Wavelength, Wavenumber, and Frequency Values for the Visible and Near-Infrared Spectral Regions

Region	Wavelength			Wavenumber (cm^{-1})	Frequency (Hertz)
	(cm)	(μm)	(nm)		
Visible	3.5×10^{-5} to 7.8×10^{-5}	0.35–0.78	350–780	28,571 to 12,821	8.563×10^{14} to 3.842×10^{14}
Extended Near-Infrared	7.8×10^{-5} to 3.0×10^{-4}	0.78–3.0	780–3000	12,821 to 3333	3.842×10^{14} to 9.989×10^{13}

Source: Workman, J., Interpretive spectroscopy for near-infrared, *Appl. Spectrosc. Revs.*, 31 (3): 251–320, 1996. With permission.

1.17 NORMAL MODE THEORY (THE IDEAL OR SIMPLE HARMONIC OSCILLATOR)

Light from a spectrophotometer is directed to strike a matrix consisting of one or more types of molecules. If the molecules do not interact with the light, then the light passes through the matrix with no interaction whatsoever. If molecules interact with the light in a very specific way (i.e., molecular absorption), we refer to them as *active* or *infrared active*. For NIR energy, the overtones of X–H bonds, i.e., N–H, C–H, and O–H stretching and bending are of the greatest interest. Infrared active molecules can be seen as consisting of mechanical models with vibrating dipoles. Each dipole model vibrates with a *specific frequency* and *amplitude* as shown using a simple model (Figure 1.12).

Note that the term *frequency* refers to the number of vibrations per unit of time, designated by the Greek letter ν (nu) and generally specified in units of sec^{-1} or Hertz (Hz). *Amplitude* is defined by the interatomic distance covered at the extremes of the vibrating dipole and is dependent upon the amount of energy absorbed by the infrared active bond. When incoming photons from a spectrophotometer source lamp (after they have passed through the monochromator or interferometer) strike different molecules in a sample, two direct results may occur: (1) the disturbing energy does not match the natural vibrational frequency of the molecule; or (2) the disturbing frequency

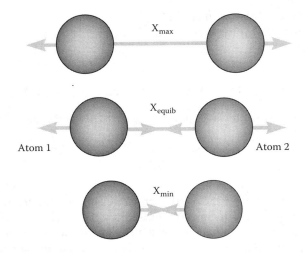

FIGURE 1.12 Model of infrared active molecule as a vibrating dipole between two atoms. (From Workman, J., Interpretive spectroscopy for near-infrared, *Appl. Spectrosc. Revs.*, 31 (3): 251–320, 1996. With permission.)

does match the vibrational frequency of the molecule. When there is a match between the disturbing frequency of the illumination energy and the natural vibrational frequency of a molecule in the sample, the molecule absorbs this energy, which in turn increases the vibrational amplitude of the absorbing dipoles. However, regardless of the increase in amplitude, the frequency of the absorbing vibration remains constant.

Another name for the dipole model from Figure 1.12 is an *ideal harmonic oscillator*. The frequency at which the dipole (or ideal harmonic oscillator) vibrates (stretches or bends) is dependent upon the bond strength and the masses of the atoms bonded together. When the harmonic oscillator (HO) vibrates, the vibrational energy is continuously changing from kinetic to potential and back again. The total energy in the bond is proportional to the frequency of the vibration. The use of Hooke's law (in our case referring to the elasticity properties of the HO) is applied to illustrate the properties of the two atoms with a well-behaved spring-like bond between them. The natural frequency of vibration for a bond (or any two masses connected by a spring) is given by the well-known relationship

$$v = \frac{1}{2\pi} \sqrt{\frac{K}{\left(\frac{1}{m_1} + \frac{1}{m_2}\right)}} \quad (1.7)$$

where K is a force constant that varies from one bond to another; m_1 = the mass of atom 1; and m_2 = the mass of atom 2. Note that as a first approximation, the force constant (K) for single bonds is 1/2 times that of a double bond and 1/3 that of a triple bond. Also note that as the mass of the atoms increases, the frequency of the vibration decreases.

Group theory can be used to represent the associations or bonds between the atoms of molecules into one-dimensional simple springs or simple harmonic oscillators. Normal mode theory, using the simple harmonic oscillator model, is able to predict with relative accuracy the energy or frequency of fundamental absorption bands, such as symmetric and asymmetric stretching, scissoring, bending, and wagging. However, for overtone band positions, normal mode theory does not predict band positions because bonds are not true harmonic oscillators. The effects of quantum mechanics on a simple HO indicate that we cannot treat the bond between two atoms quite as simply as two masses connected by a spring. This is no surprise because quantum mechanical evidence has shown that vibrational energy between atoms in a molecule is quantized into discrete energy levels. When the conditions are right, vibrational energy in a molecule "jumps" from one energy level to another. The discrete vibrational energy levels for any molecule E_{VIB} are given by

$$E_{VIB} = hv\left(v + \frac{1}{2}\right) \quad (1.8)$$

where h = Planck's constant; v (Greek nu) = vibrational frequency of the bond; and v = the vibrational quantum number (which can only have the integer values of 0, 1, 2, 3, and so on).

1.18 THE ANHARMONIC OSCILLATOR

The concept of an *anharmonic oscillator* allows for more realistic calculations of the positions of the allowed overtone transitions. The energy levels of these overtones are not found to be the product of the exact integer multiplied by the fundamental frequency. In fact, the following expression defines the relationships between wavenumber (for a given bond) and the vibrational energy of that bond using local mode or anharmonicity theory. The relationship is calculated from the Schroedinger equation to yield

$$\overline{v} = \left(\frac{E_{VIB}}{hc}\right) = \overline{v}_1 v - x_1 \overline{v}_1 (v + v^2) \tag{1.9}$$

where v = an integer number, i.e., 0, 1, 2, 3,..., n, and $x_1 \overline{v}_1$ = the unique anharmoni-city constant for each bond.

Calculations of band positions using Equation 1.18 will more closely approximate observed band positions than those calculated from the ideal harmonic oscillator expression found in Equation 1.16. For a rule of thumb, the first overtone ($2v$) for a fundamental can be calculated as 1% shift due to anharmonicity, or x_1 = (0.01). Thus, the expression using wavenumbers is

$$\overline{v} = \overline{v}_1 v - \overline{v}_1 (0.01) \cdot (v + v^2) \tag{1.10}$$

To illustrate the occurrence of free O–H first overtone absorption using a fundamental absorption occurring at approximately 3625 cm^{-1}, the first overtone (in wavenumbers) of free O–H stretching should occur at

$$\overline{v} = (3625 \cdot 2) - 3625(0.01) \cdot (2 + 2^2) = 7033 \text{ cm}^{-1} \text{ or } 1422 \text{ nm} \tag{1.11}$$

Thus, one would expect the first overtone to occur somewhere near 7033 cm^{-1} or 1422 nm using a simple harmonic oscillator model (to convert to wavelength, use $10^7 \div v$ [in cm^{-1}]. Calculations for wavenumber positions for the first overtone ($2v$), second overtone ($3v$), and third overtone ($4v$) can be estimated with the assumption of a 1–5% frequency shift due to anharmonicity.

Molecules that absorb NIR energy vibrate primarily in two fundamental modes: (1) *stretching* and (2) *bending*. Stretching is defined as a continuous change in the interatomic distance along the axis between two atoms, and bending is defined as a change in the bond angle between two atoms. Figure 1.13 illustrates the often repeated stretching and bending interactions defining infrared active species within infrared active molecules.

Note that near-infrared (12,821–4000 cm^{-1}, 780–2500 nm) spectral features arise from the molecular absorptions of the overtones and combination bands that originate from fundamental vibrational bands generally found in the mid-infrared region. For fundamental vibrations, there exists a series of overtones with decreasing intensity as the transition number (overtone) increases. Combination bands arise as the summation of fundamental bands, their intensity decreasing with an increase in the summation frequency. Most near-infrared absorptions result from the harmonics and overtones of X–H fundamental stretching and bending vibrational modes. Other functional groups relative to near-infrared spectroscopy can include hydrogen bonding, carbonyl carbon to oxygen stretch, carbon to nitrogen stretch, carbon to carbon stretch, and metal halides.

Molecular vibrations in the near-infrared regions consist of stretching and bending combination and overtone bands. Stretching vibrations occur at higher frequencies (lower wavelengths) than bending vibrations. Stretching vibrations are either symmetric or asymmetric; bending vibrations are either in-plane or out-of-plane. In-plane bending consists of scissoring and rocking; out-of-plane bending consists of wagging and twisting. From the highest frequency to lowest, the vibrational modes occur as stretching, in-plane bending (scissoring), out-of-plane bending (wagging), twisting, and rocking. The most-often observed bands in the near-infrared

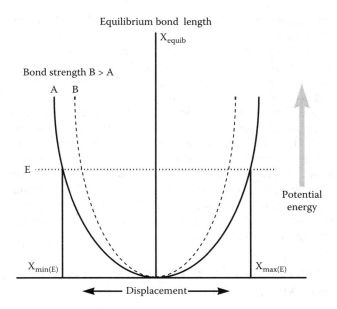

FIGURE 1.13 Potential energy curve showing minimum and maximum amplitude for the harmonic oscillator model as a continuum. (From Workman, J., Interpretive spectroscopy for near-infrared, *Appl. Spectrosc. Revs.*, 31 (3): 251–320, 1996. With permission.)

include the combination bands and first, second, or third overtones of O–H, N–H, and C–H fundamentals.

Variations in hydrogen bonding manifest themselves as changes in the force constants of the X–H bonds. Generally, bands will shift in frequency and broaden due to the formation of hydrogen bonding. Because combination bands result from the summation of two or more fundamental vibrations, and overtones occur as the result of the multiples of fundamental vibrations, frequency shifts related to hydrogen bonding have a greater relative effect on combination and overtone bands than on their corresponding fundamentals. This feature of the near-infrared region alerts one to the importance of the relative hydrogen bonding effects brought about by solvent and temperature variations.

Precise band assignments are difficult in the near-infrared region because a single band may be attributable to several possible combinations of fundamental and overtone vibrations, all severely overlapped. The influence of hydrogen bonding results in band shifts to lower frequencies (higher wavelengths); a decrease in hydrogen bonding due to dilution and higher temperatures results in band shifts to higher frequencies (lower wavelengths). Band shifts of the magnitude of 10–100 cm^{-1}, corresponding to a few to 50 nm, may be observed. The substantial effect of hydrogen bonding should be kept in mind when composing calibration sample sets and experimental designs for near-infrared experiments.

Near-infrared spectra contain information relating to differences in bond strengths, chemical species, electronegativity, and hydrogen bonding. For solid samples, information with respect to scattering, diffuse reflection, specular reflection, surface gloss, refractive index, and polarization of reflected light are all superimposed on the near-infrared vibrational information. Aspects related to hydrogen bonding and hydronium ion concentration are included within the spectra.

Light can interact with the sample as reflection, refraction, absorption, scattering, diffraction, and transmission. Signal losses from the sample can occur as specular reflection, internal scattering, refraction, complete absorption, transmission loss during reflectance measurements, and trapping losses. Spectral artifacts can also arise as offset or multiplicative errors due to coloration of the sample, variable particle sizes and resultant variability in apparent pathlength, refractive index changes in clear liquids relative to temperature changes, and pathlength differences due to temperature-induced density changes.

During measurement of a sample, the light energy entering the sample will be attenuated to some extent; the light entering the sample is able to interact with the sample, emerging as attenuated transmitted or reflected light. The frequency and quantity of light absorbed yields information regarding both the physical and compositional information of the sample. For reflection spectra, the diffuse and specular energy fractions are superimposed. The intensity of reflected energy is a function of multiple factors, including the angles of incidence and observation, the sample packing density, crystalline structure, refractive index, particle size and distribution, and absorptive and scattering qualities.

The types of vibrations found in near-infrared and infrared spectroscopy are designated by ν (Greek letter nu) with a subscript designating whether the vibration is symmetric or asymmetric and bending by δ. Combination bands resulting from the sum of stretching and bending modes are designated as $\nu + \delta$; harmonics are designated as $k\nu$ where k = an integer number as 2 (first overtone), 3 (second overtone),..., k (k-1 overtone), and ν is the frequency of the fundamental stretch vibration for a specific functionality.

1.19   ILLUSTRATION OF THE SIMPLE HARMONIC OSCILLATOR OR NORMAL MODE

The classical harmonic oscillator model or normal mode is often illustrated using a potential energy curve (Figure 1.14). The larger the mass, the lower the frequency of the molecular vibration, yet the potential energy curve does not change. The relationship for a simple diatomic molecule model is

$$Frequency = \nu(nu) = \frac{1}{2\pi}\sqrt{\frac{k}{\mu}} \qquad (1.12)$$

where k is the force constant for the bond holding the two atoms together, and μ is the reduced mass from the individual atoms.

For any bond, there is a limit to the amplitude of motion away from and toward the opposing atom (X_{max} and X_{min}, respectively); the potential energy increases at either extreme, i.e., either the minimum or maximum distance between atoms. The total potential energy (E_τ) for a bond following the harmonic oscillator model is given by $E_\tau = \frac{1}{2}k \cdot x^2$, where k is the force constant for the bond holding the two atoms together, and x is the change in displacement from the center of equilibrium. For any specific bond, there is a limit to the total energy and maximum displacement of the atoms (or dipoles). Quantum mechanical theory is used to more closely describe the behavior of these atoms connected by bonds with varying force constants. Using quantum theory, it is possible to describe a mathematical relationship for the behavior of two oscillating atoms as following not a continuous change in potential energy states for the vibrating system but actually a set of discrete energy levels E_n given by $E_n = (n + \frac{1}{2})h\nu$, where n is the oscillating dipole quantum number for discrete potential energy levels , i.e., 0, 1, 2, 3, ..., n. The potential energy curve better describing the behavior of oscillating dipoles is given in Figure 1.14.

1.20   THE SELECTION RULE

The selection rule states that, in the quantum mechanical model, a molecule may only absorb (or emit) the light of an energy equal to the spacing between two levels. For the harmonic oscillator (dipole involving two atoms), the transitions can only occur from one level to the next higher (or lower) level.

The infrared spectrum for any molecule will exhibit a band corresponding to the frequency of the energy at which light can absorb, namely, E_0, E_1, E_3, and so on. Figure 1.14 illustrates the discrete

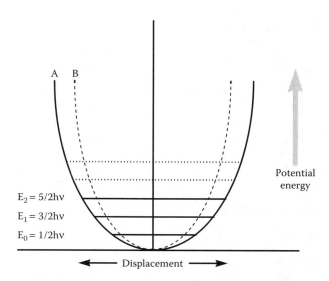

FIGURE 1.14 Illustration of the discrete quanta potential energy curves (E_0, E_1, E_2, and so on.) for both strong and weakly bonded dipoles as $E_n = (n + \frac{1}{2})hv$. (From Workman, J., Interpretive spectroscopy for near-infrared, *Appl. Spectrosc. Revs.*, 31 (3): 251–320, 1996. With permission.)

quanta potential energy curves for both strongly and weakly bonded dipoles. The energy levels are not equally spaced for the harmonic oscillator as shown in the following text. Note that the figure illustrates the energy levels for the fundamental vibrations (or oscillations) only. In addition to the fundamentals, the overtone bands also occur at less than two times, three times, and four times the frequency of the fundamental vibrations, and at much less intensity; further discussions on this will describe the factors involving the infrared and near-infrared spectroscopic activity of molecules.

1.21 ILLUSTRATION OF THE ANHARMONIC OSCILLATOR

The model of the anharmonic oscillator or local mode approximation more closely follows the actual condition for molecular absorption than that of the harmonic oscillator. Figure 1.15 illustrates the differences between the ideal harmonic oscillator case that has been discussed in detail for this chapter vs. the anharmonic oscillator model (better representing the actual condition of molecules). Unlike the ideal model illustrated by the harmonic oscillator expression, the anharmonic oscillator involves considerations such that, when two atoms are in close proximity (minimum distance), they repel each other; when two atoms are separated by too large a distance, the bond breaks. The anharmonic oscillator potential energy curve is most useful to predict behavior of real molecules.

1.22 INTERPRETIVE NEAR-INFRARED SPECTROSCOPY

The energy absorbed by a matrix consisting of organic compounds depends upon the chemical composition of the matrix, defined by the species (or type) of molecule present, the concentration of these individual species, and the interactions between the molecules in the matrix. In order for NIR (or any other vibrational spectroscopic measurement technique) to be valid, one must be absolutely assured that different types of molecules absorb at unique frequencies. Due to the broadband nature of NIR spectra consisting of overlapping combination and overtone bands, the individual species are not well resolved as they are in the mid-infrared region. In addition, many compounds absorb NIR energy throughout the entire wavelength region, making it difficult, if not

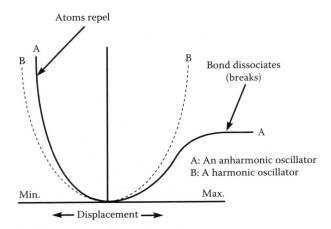

FIGURE 1.15 Illustration of the differences in potential energy curves between the ideal harmonic oscillator model and the harmonic oscillator model (better representing the actual condition of molecules). (From Workman, J., Interpretive spectroscopy for near-infrared, *Appl. Spectrosc. Revs.*, 31 (3): 251–320, 1996. With permission.)

impossible, to clearly resolve a usable baseline for simple peak-height or peak-area quantitative methods. This brings one to the same conclusion drawn by early investigators that the NIR region is not especially useful as a quantitative measurement technique. At the very least, novel techniques for spectral manipulation were required to "interpret" the poorly resolved bands and "compensate" for background interferences.

1.23 GROUP FREQUENCIES

Frequencies that are characteristics of groups of atoms (termed *functional groups*) are given common group names and can be assigned general near-infrared locations (in nanometers) for major groups associated with stretching vibrational bands. The locations of the fundamental frequencies, overtones, and combination bands are provided in this book. References 5 and 6 provide a rich source of information as to locations of fundamental frequencies for the major function groups (see Table 1.5 for examples) within the infrared and near-infrared regions.

1.24 COUPLING OF VIBRATIONS

Coupling indicates that the oscillators or molecular vibrations of two or more molecules are interactive, so that the original vibrational energy states (if the vibrations could occur independently of one another) result in split energy states due to the interaction of the vibrations. Coupling is divided into two basic orders: first and second (Fermi resonance). First-order coupling can be involved in several important infrared group frequencies. For example, CO_2 has two separate uncoupled oscillators of C=O, each occurring at approximately 5405 nm (v at ~1850 cm^{-1}), 2756.6–2972.8 nm ($2v$), 1855.7–2071.9 nm ($3v$), and 1405.3–1621.5 nm ($4v$). The interactive (coupled oscillators) energy states occur at 4257 nm (v at 2349 cm^{-1}, asymmetric stretch), 2171.1–2341.4 nm ($2v$), 1461.6–1631.9 nm ($3v$), and 1106.8–1277.1 nm ($4v$); and 7463 nm (v at 1340 cm^{-1}, symmetric stretch), 3806.1–4104.7 nm ($2v$), 2562.3–2860.8 nm ($3v$), and 1940.4–2238.9 nm ($4v$). First-order coupling involves multiple infrared group frequencies including:[6]

TABLE 1.5
Example Functional Group Names and General Formulas in *Simplified Molecular Input Line Entry Specification* or SMILES Format

Hydrocarbons, etc.

C–C, paraffinics (or alkanes)
$CH_3-(CH_2)_N-CH_3$, normal paraffins
R–CH_3, methyl C–H
R–CH_2 R′, methylene C–H
R–CHR′R″, methine C–H
C1CCCCC1, cyclohexane
>C=C<, olefinic group (or alkenes)
–CH=CH_2, vinyls
R′R″C=CH_2, vinylidenes
>C=C=CH_2, allenes
–C≡C–acetylenes (or acetylinics)
C_NH_N, aromatics
C_NH_{2N}, naphthalenes (or cycloalkanes)

Carbonyl Compounds

R–C(=O)–H, aldehydes
R–C(=O)–R′, ketones
>C=C<HC(=O)–R′, ketenes
R–C(=O)–OR′, esters
S–C(=O)–O–R′, thiol esters
R–C(=O)–C(=O) –R, anhydrides
–C–O–O–C–, peroxides
–O–C(=O) –C(=O) –O–, oxalates
–O–C(=O) –, carboxy–
–C(=O) –, carbonyl group
–C(=O) –NH_2, primary amides
–C(=O) –NHR, secondary amides
–C(=O) –NR′R″, tertiary amides

Ethers

–OCH_3, methoxy (or ether group)
C–O–C, ethers

Hydroxy

–OH, hydoxyl–[O–H]
R–OH, alcohols [O–H]
–CH_2–OH, primary alcohols:
R′R″CH–OH, secondary alcohols
R′R″R‴C–OH, tertiary alcohols
Ar–OH, phenolics (or phenols) [O-H]
Ar=c1ccccc1.

Nitrogen Compounds

–C≡N, or C#N, nitriles
R–C≡N–, or R–C#N–, nitriles
S–C≡N⁻, or S– C#N⁻, thiocyanate
C≡N⁻, or C#N, cyanide
C–NO, nitroso group
R′R″NNO, nitrosamines
R–NO_2^-, nitro group (or nitrite)
R–NH_2, primary amines
R′R″NH, secondary amines
R–NH_3^+, R′R″NH_2^+, amine salts
N=N, azo group
–N=N⁺=N⁻, –N⁻–N⁺≡N; azides
NNO, azoxy group
R–O–N–N=O, organic nitrites
–NO_3, nitrates
ON=NO, nitroso- group
H_3N^+–CH–C(=O)O⁻, amino acids

The stretches for all cumulated double bonds, X=Y=Z, e.g., C=C=N
The stretches in XY_2, including –CH_2–, and H_2O
The stretches in XY_3 groups, including –CH_3
The deformations of XY_3 groups, including –CH_3, CCl_3
The N–H in-plane bend of secondary amides, e.g., R–CO–NH–R′

1.25 FERMI RESONANCE (OR SECOND-ORDER COUPLING)

Fermi resonance[8] is the interaction or coupling of two vibrational energy states with the resultant separation of the states where one of them is an overtone or a sum tone. An overtone vibration occurs as the integer multiple of a fundamental vibration in frequency space (intensity falls off rapidly with the higher multiples):

The first overtone of a fundamental vibration (v) is equal to $2 \times v$.
The second overtone of a fundamental vibration is equal to $3 \times v$.
The third overtone is equal to $4 \times v$.

Note: Overtones are a special case of sum tones where the frequencies are identical. A sum tone is the general case of an overtone where the frequencies are not equal and where a variety of vibrational energy states can occur.

A binary sum tone is equal to the sum of two fundamentals, e.g., $v_i + v_k$.
A ternary sum tone is equal to the sum of three fundamental vibrations, e.g.,

$$v_i + v_k + v_m.$$

Other sum tones can occur such as the sum of an overtone and a fundamental vibration,
  e.g., $2v_i + v_k$.

Three requirements are stated for Fermi resonance:[6]

1. The zero-order frequencies must be close together (typically within 30 cm^{-1}).
2. The fundamental and the overtone or sum tone must have the same symmetry.
3. There must be a mechanism for the interaction of the vibrations.

Note: (1) The vibrations cannot be separated (or localized) in distinctly different parts of the molecule. (2) The vibrations must be mechanically interactive in order that the interaction of one vibration affects another.

The results of Fermi resonance are important for infrared and near-infrared spectroscopy. Fermi resonance causes the following effects on spectral bands:

The resultant bands are moved in position from their expected frequencies.
Overtone bands are more intense than expected.
There may exist doublet bands where only singlets were expected.
Solvent changes can bring about slight shifts in frequency location of a band, and intensities
  can be greatly changed.

1.26 TOOLS AND TECHNIQUES FOR ASSIGNING BAND LOCATIONS

There are many current techniques used for assisting in near-infrared band assignments, including Darling–Dennison[9] resonance, deuteration, polarization, and two-dimensional (2-D) correlation spectroscopy. The scope of this book is practical rather than theoretical, and so the techniques are mentioned here for further investigation by the reader.

For some molecules, for example H_2O, the near-resonant frequencies of the symmetric (s) and asymmetric (a) stretches interact in such a way that the normal stretching modes in the molecule couple. Addition of the Darling–Dennison coupling in a molecule gives a band that is no longer separable into the two s and a modes. The Darling–Dennison coupling is for two modes s and a, which have approximately the same frequency, which is often called a 1:1 resonance coupling.

Heavy hydrogen or deuterium is used to deuterate compounds and correlate the changes in their near-infrared spectral band positions, intensities, and widths for both the deuterated (deuterium-labeled) and undeuterated (unlabeled) molecules. This experiment provides information for the identification of the vibrational modes of the deuterated molecule based on the known or accepted assignments of the nondeuterated molecular forms.

Polarization and dichroism methods, such as variable circular dichroism (VCD) and magnetic circular dichroism (MCD), are used to determine band assignments in complex molecules. By using various models, polarization features can be correlated with the observed positions of bands in the near-infrared spectra. Absorption bands are assigned to short-axis polarized Q transitions measured using such techniques.

Two-dimensional correlation spectroscopy is used for detailed band assignment work. The technique allows spectral information to be analyzed that is much richer in information content than one-dimensional data. Cross-correlation analysis methods are applied to spectral combinations of NIR with NIR, or NIR and mid-infrared, allowing band assignments to be more easily accomplished. An excellent review paper describing the mathematics used in 2-D correlation spectroscopy along with several examples of generalized 2-D NIR and 2-D NIR-mid-infrared (MIR) heterospectral correlation analysis are introduced with 42 references by Ozaki and Wang.[10]

REFERENCES

1. Del Fanti, N.A., Bradley, K., Bradley, M., Izzia, F., and Workman, J., *IR Spectroscopy* of Polymers, ThermoFisher Scientific Corp., 2007, p. 162.
2. Colthup, N.B., Daly, L.H., and Wiberley, S.E., *Introduction to Infrared and Raman Spectroscopy*, Academic Press, Boston, MA, 1990, pp. 119–213, 483–542.
3. Walton, P.H., *Beginning Group Theory for Chemistry*, Oxford University Press, Oxford, 1998.
4. Workman, J. and Coates, J., Interpretive Spectroscopy for SW-NIR, *The Pittsburgh Conference*, No. 126, 1995.
5. Goddu, R.F. and Delker, D.A., Spectra-structure correlations for near-infrared, *Anal. Chem.*, 32, 140–141, 1960.
6. Mayo, D.W. (Ed.), *Infrared Spectroscopy: I. Instrumentation; II. Instrumentation, Raman Spectra, Polymer Spectra, Sample Handling, and Computer-Assisted Spectroscopy* (Vol. 1), Bowdoin College, ME, 1994.
7. Hershel, W.J., *Nature,* 23, 76, November 25, 1880.
8. Fermi, V.E., Uber den Ramaneffekt des Kohlendioxyds, *Z. Phys.,* 71, 250–259, 1931.
9. Darling, B.T. and Dennison, D.M., *Phys. Rev.,* 57, 128, 1940.
10. Ozaki, Y. and Wang, Y., Two-dimensional near-infrared correlation spectroscopy: principle and its applications, *J. Near-Infrared Spectrosc.*, 6(1–4), 19–31, 1998.

2 Alkanes and Cycloalkanes

2.1 C-H FUNCTIONAL GROUPS

The fact that aliphatic hydrocarbons had an absorption band at 1.7 μm was known as early as 1881.[1] The assignment of this absorption as a first overtone of a C–H stretch vibration was perhaps first recognized by Joseph Ellis, although he initially felt that the 1.7-μm peak was part of a harmonic series that began with the 6.9-μm CH bending mode.[2] Nevertheless, the assignment of the first overtone of carbon–hydrogen stretching vibrations was made very early in the history of near-infrared (IR) spectroscopy.

Aliphatic and aromatic C–H-stretching first overtones are rich in information content, having much the same information as the fundamental vibrations. In addition, the combinations of these stretches with other vibrational modes provide much of the spectral structure of the near-infrared region.

As a general assignment, the first overtones of C–H stretching occur between 5555 and 5882 cm^{-1} (1700–1800 nm), the second overtones between 8264 and 8696 cm^{-1} (1150–1210 nm), and the third overtones between 11,364 and 10,929 cm^{-1} (880–915 nm). The most important combination regions occur between about 6666 and 7690 cm^{-1} (1300–1500 nm) and 4545 and 4500 cm^{-1} (2200–2500 nm).

2.2 METHYL GROUPS, CH$_3$

Figure 2.1 shows an example with some general band assignments of the methyl C–H spectral regions for iso-octane (2,2,4-dimethyl pentane), a molecule that has mostly methyl groups.

2.2.1 FIRST OVERTONE REGION

As seen in Figure 2.2, a homologous series of alkanes has four main peaks in the first overtone region. In the mid-infrared fundamental region, these peaks represent the methyl asymmetric, methylene asymmetric, methyl symmetric, and methylene symmetric vibrations in the order from the highest to the lowest wavenumber. In the near-infrared, however, these peaks do not fall in this sequence, and their actual assignments are more complex. As shown in Figure 2.2, the two lower wavelength (higher wavenumber) peaks decrease as the chain length increases, clearly indicating that these two belong to the methyl group. These peaks have been referred to as the first overtones of the asymmetric and symmetric stretch vibrations, but this may be an oversimplification.

As described by Tosi and Pinto,[3] who examined 50 linear and branched hydrocarbons, there are five primary bands in the first overtone region. The first appears at 5905 +/– 4 cm^{-1} in all of the hydrocarbons and can be attributed to the methyl group. It is probably the first overtone of the asymmetric stretch of the methyl group. The position of the second peak decreases in wavenumber with an increase in chain length and follows the formula 5853 + 49F–CH$_3$, where F–CH$_3$ is the mole fraction of methyl groups in a given hydrocarbon. For hexane, for example, this peak would be estimated to be at 5876 cm^{-1}. As seen in Figure 2.2, the methyl peaks of hexane are at 5907 and 5870 cm^{-1}. The second peak appears to be initially stronger but becomes less distinct when the chain length is increased as it becomes affected by the neighboring methylene peak.

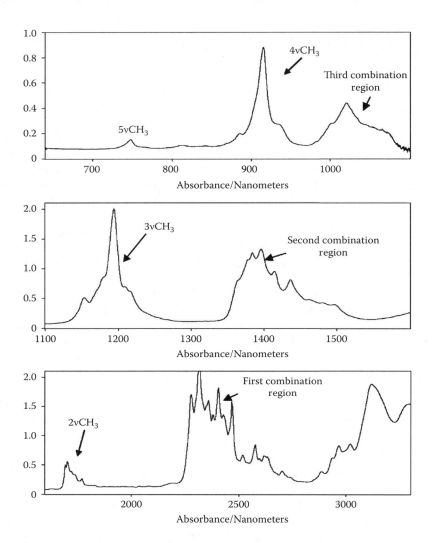

FIGURE 2.1 Illustrates aliphatic C−H (as predominantly methyl C−H str.) from 2,2,4-dimethyl pentane (iso-octane). The top spectrum (from left to right) shows the 4th overtone (5ν), the 3rd overtone (4ν), and combination bands. The center spectrum illustrates the 2nd overtone (3ν) and combination bands. The bottom spectrum presents the 1st overtone (2ν) and combination bands, and a portion of the fundamental. (From Workman, J., *Appl. Spectrosc. Revs.*, 31(3), 277, 1996. With permission.)

Within the classes of hydrocarbon compounds (linear, singly branched, doubly branched), good correlations between the peak intensities and the percent of the functional group can be obtained. In general, the absorptivity of terminal methyls is higher than that of internal or branched methyls.

As in the mid-infrared, the position of the C−H stretch vibrations can be affected by adjacent atoms. Table 2.1 lists some examples. Figure 2.3 illustrates the shift of the two methyl "overtone" peaks with the influence from the carbonyl group.

In addition to the two methyl peaks near 5790 and 5735 cm^{-1}, the methyl group in aromatic compounds has a band near 5660 cm^{-1}, which has been found useful for quantitative analysis. Luty and Rohleder[4] assigned this peak to be due to $\nu_a + 2\delta_a$. They also suggest that a peak at 4080 cm^{-1} is due to $3\delta_s$, the second overtone of a symmetric bending vibration of the CH_3 group. This peak is well isolated in compounds having multiple methyl groups such as penta and hexa-methyl benzene.

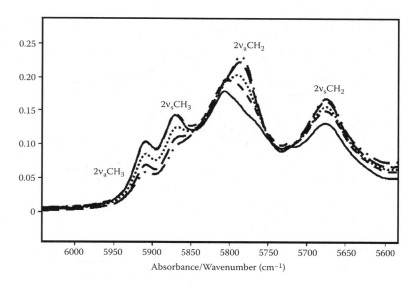

FIGURE 2.2 First overtones of methyl and methylene C–H stretching; solid line is hexane, small dash heptane, dots nonane, long dash dodecane, and dot-dash hexadecane. (From Weyer, L.G. and Lo, S.–C., in *Handbook of Vibrational Spectroscopy,* Chalmers, J.C. and Griffiths, P.R., John Wiley & Sons, 2002. With permission.)

TABLE 2.1
Peak Positions for the Methyl Group in the 5000–6000 cm⁻¹ Region

Aliphatic, terminal R–CH₃ terminal   5905; 5876

Aliphatic, branched   CH₃   5905; 5872
                              |
                         –R–CH–R–

Aromatic, Ar   5790; 5735; plus 5650

Carbonyl   O   5960; 5898
                  ‖
              –C–CH₃

Carbonyl, one carbon   5946; 5908
removed   O
                  ‖
              –C–CH₂CH₃

Hydroxyl   HO–CH₃   5880; 5773

Ether   CH₃–O–CH₃   5880; 5770

Amine   R–C–NH₂

Halogen   CH₃Cl   6000/6040 doublet, 5882 — gas
              CH₃Br   6025/6060 doublet, 5900 — gas
              CH₃I   6020; 5902; 5845

Nitro   6045; 5874

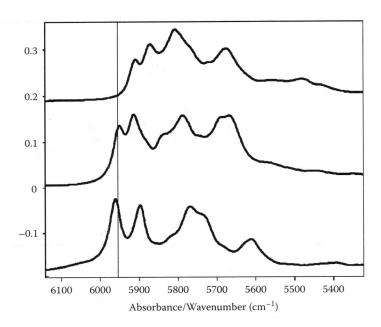

FIGURE 2.3 Shift of methyl overtone peaks; top spectrum is hexane, middle is 3-pentanone, and bottom is acetone.

2.2.2 HIGHER-ORDER OVERTONES

As seen in Figure 2.1, the second, third, and fourth overtone regions (3v, 4v, and 5v) have only one strong band. This was thought to be most likely due to the asymmetrical vibrations, as the symmetrical bands became relatively weaker in higher overtones.[5] An alternate theory for this phenomenon is that local mode effects at the higher overtones combine to create only one general absorption peak. The second overtone is generally considered intermediate between the normal mode and local mode, whereas the higher-order overtones are only local mode.

The second overtone region (1150–1210 nm or 8264–8696 cm⁻¹) has also been used for quantitative measurements, in particular to measure methyl, methylene, methine, and aromatic contributions.[6] The methyl groups of long-chain paraffinic hydrocarbons appear between 8365 and 8375 cm⁻¹ (1194–1195nm).[7] In pentane and hexane, the methyl group absorbs at 8396 cm⁻¹ (1191 nm), in heptane it absorbs at 8388 cm⁻¹ (1192 nm), and in decane it is at 8378 cm⁻¹ (1194 nm). See Figure 2.4.

The third overtone vibration of the methyl group appears at 10,953 in hexane. The methyl third overtone in some additional molecules is listed in Table 2.2.[8] Wheeler[9] assigns the fourth overtone's position to be at about 13,400 cm⁻¹ (746 nm). Fang and Swofford[10] list the fifth overtone for a linear alkane methyl group C–H stretch to be at 15,690 cm⁻¹ (637 nm), and the sixth at about 17,890 cm⁻¹ (560 nm).

2.2.3 COMBINATION BANDS

The first "combination" region near 4500–4545 cm⁻¹ (2200–2500 nm) is rich in information but complex, as it includes combinations of C–H stretching with various bending and stretching

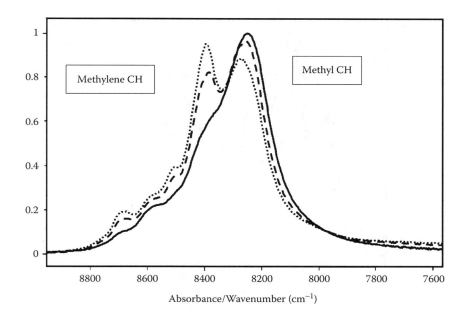

FIGURE 2.4 The second overtone region, illustrating the separation of methyl and methylene contributions; the figure shows hexane (dotted curve), octane (dashed), and octadecane (solid).

vibrations from the mid-infrared "fingerprint" region. In addition, overtones of bending modes alone may be present.

For the methyl functional group, there is a strong peak near 4395 cm⁻¹ (2275 nm), as seen in Figure 2.5. This is probably a $\nu + \delta$ combination. There is also said to be a 3δ peak of the symmetric methyl bending "umbrella" vibration near 4100 cm⁻¹ (2410–2460 nm).[11,12] A 3δ overtone of the asymmetric bending mode is expected to be at 4400 cm⁻¹ (2270 nm). A third overtone of the symmetric bending vibration, 4δ, is cited at about 5520 cm⁻¹ (1812 nm), whereas a third overtone of the asymmetric bending vibration is suggested to be at 5814 cm⁻¹ (1720 nm).

TABLE 2.2
Example Third Overtone Methyl Peaks

Compound	Peak Position (cm⁻¹)	Peak Position (nm)
1-Bromoalkane	10,953	912
1-Heptene	10,953	912
n-Hexane	10,953	912
1-Amino hexane	10,941	914
2-Amino propane	10,977	911
Propionaldehyde	11,052	905
3-chloro-2-methyl-1-propane	10,965	912
1-Octanol	10,953	912
Toluene	10,977	911
Ethylbenzene	10,953	912

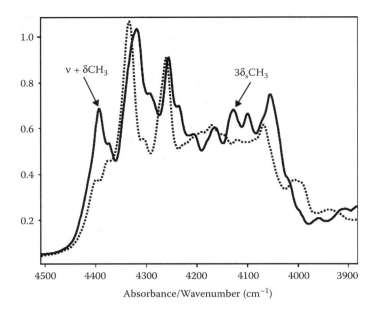

FIGURE 2.5 First combination region of 2,5-dimethyl hexane (solid line) and decane (dashed line).

In the 7100-cm^{-1} (1400-nm) region of second combination bands, the double peak near 7355 cm^{-1} (1360 nm) and 7263 cm^{-1} (1377 nm) is suggested to be a combination of $2\nu + \delta$.[13] Also, in branched dimethyl alkanes, there is another, larger peak at 6938 cm^{-1} (1440 nm). This is shown in Figure 2.6.

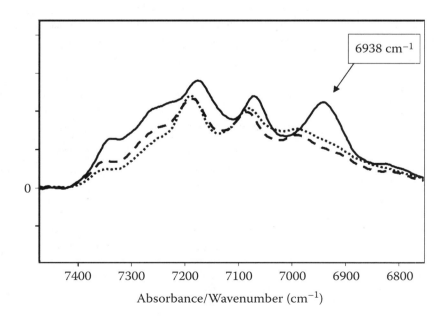

FIGURE 2.6 Illustration of branched hydrocarbon second combination band; 2,5-dimethyl hexane (solid line), hexane (dashed line), decane (dotted line).

2.3 METHYLENE GROUPS, CH$_2$

The general absorptions of methylene groups are illustrated in Figure 2.7, a spectrum of *n*-decane.

2.3.1 FIRST OVERTONE REGION — LINEAR MOLECULES

Methylene groups in linear, aliphatic molecules have two primary peaks at about 5800 cm^{-1} (1723 nm) and 5680 cm^{-1} (1762 nm) in the first overtone region. The 5800-cm^{-1} peak is generally thought to be a combination band,[14] reported as $v_a + v_s$.[15] The 5680-cm^{-1} peak is considered to be a pure first overtone of the asymmetric stretch (Bubeck) or the symmetric stretch (Ricard-Lespade, Durbetaki). For solids, Bubeck lists both the asymmetric and symmetric stretches.

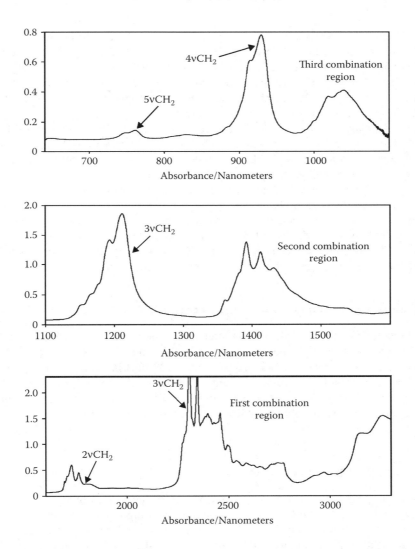

FIGURE 2.7 Illustration of Aliphatic C–H (as predominantly methylene C–H structure) from *n*-decane. The top spectrum (from left to right) shows the fourth overtone (5v), the third overtone (4v), and combination bands. The center spectrum illustrates the second overtone (3v) and combination bands. The bottom spectrum presents the first overtone (2v) and combination bands, and a portion of the fundamental. (From Workman, J., *Appl. Spectrosc. Revs.*, 31(3) 277, 1996. With permission.)

TABLE 2.3
Example Third Overtone Methylene Peaks

Compound	Peak Position (cm⁻¹)	Peak Position (nm)
1-Bromoalkane	10,764	929
1-Heptene	10,787	927
n-Hexane	10,776	928
Propionaldehyde	10,787	927
1-Octanol	10,776	928
Ethyl benzene	10,753	930
1-Amino-hexane	10,753	930

The 5800-cm⁻¹ peak is usually the strongest peak in the first overtone region of a series of linear hydrocarbons. Tosi and Pinto provide a formula for locating this peak for a series of linear hydrocarbons: 5856–85 × weight-fraction of CH_2, or about 5800 cm⁻¹ for hexane. They also mention that this peak splits into two closely spaced peaks possibly due to the influence of adjacent methyl groups. The absorptivity of this combined peak does not regularly increase with chain length, probably because it has contributions from two sources.

The weaker methylene peak in the first overtone region was said to be at 5671 +/– 3 cm⁻¹ for linear hydrocarbons. This peak was thought to have contributions from both methyl and methylene groups (Tosi and Pinto), although others have assigned a 5680-cm⁻¹ band to be both the first overtone of the methylene symmetric stretch and of the asymmetric stretch shifted by Fermi resonance (Ricard-Lespade et al.). More recently, Parker et al. have discussed the origins of this peak in terms of local mode theory.

In the second overtones, which are also shown in Figure 2.6, only one methyl and one methylene peak are normally observed at 8389 cm⁻¹ (1192 nm) and 8264 cm⁻¹ (1210 nm), although weaker peaks can be seen at 8673 cm⁻¹ (1153 nm) and 8503 cm⁻¹ (1176 nm) with higher resolution. In higher alkanes above dodecane, the methyl group becomes a shoulder in the methylene peak.[16]

2.3.2 HIGHER-ORDER OVERTONES — LINEAR MOLECULES

The second overtone of the methylene C–H stretch absorption occurs at 8284 cm⁻¹ (1207 nm) in pentane, 8271 cm⁻¹ (1209 nm) in hexane, 8256 cm⁻¹ (1211 nm) in heptane, and 8247 cm⁻¹ (1212 nm) in decane. The third overtone is at 10,776 cm⁻¹ (928 nm) in hexane. The peak positions of some additional compounds are provided in Table 2.3 from Salzer et al. The fourth overtone is at about 13,100 cm⁻¹ (762 nm). Fang and Swofford also list the fifth overtone for a linear alkane methylene group C–H stretch to be at 15,400 cm⁻¹ (649 nm), and the sixth at about 17,535 cm⁻¹ (570 nm).[17]

2.3.3 COMBINATION BANDS, LINEAR MOLECULES

The two largest peaks in the first combination region seen in Figure 2.7 are due to the asymmetric and symmetric methylene stretching and bending combinations.[18] The asymmetric combination is at 4336 cm⁻¹ in butane, 4334 in pentane, and 4332 in heptane, whereas the symmetric one is at 4257 in butane, 4262 in pentane, and 4259 in heptane, according to Ricard-Lespade et al.

In an examination of the spectrum of methylene chloride as interpreted by Kaye,[19] the methylene combination peaks at 3945, 4196, 4253, and 4453 cm⁻¹ in CH_2Cl_2 are explained as $\nu + \delta$ combinations as listed in Table 2.4. The analogous bands in a normal hydrocarbon would be at about 4068, 4168, 4261, and 4333 cm⁻¹.

TABLE 2.4
First Combination Bands of Methylene Groups in CH$_2$Cl$_2$

Assignment	Peak Observed (cm^{-1})	Peak Calculated	Summation
$v_a + \delta_r$	3945	3951	3055 + 896
$v_a + \delta_t$	4196	4211	3055 + 1156
$v_s + \delta_w$	4253	4251	2987 + 1264
$v_a + \delta_b$	4453	4477	3055 + 1422

Note: "r" denotes a rocking vibration, "t" twisting, "w" wagging, and "b" bending.

In the second combination region, shown in Figure 2.7, there is a doublet relating to the methylene group, the larger peak, at about 7186 cm^{-1} (1391 nm), and the smaller at about 7080 cm^{-1} (1412 nm). This doublet has been attributed to $2v + \delta$.[20] The $2v$ term is neither asymmetric nor symmetric, as the vibrations are uncoupled at this point. Kaye called it $v_{a+}v_s$. Also, δ would include rocking, twisting, and/or bending vibrations. The two strongest bands are probably $2v + \delta_b$ and $2v + \delta_t$, based on Kaye's description of the spectrum of methylene chloride.

There are also a number of weaker combination bands in the area between the first combination and the first overtones regions, 4500–5500 cm^{-1} or about 1800 to 2220 nm. These include $v_a + 2\delta_r$, $v_s + \delta_r + \delta_t$, $v_s + 2\delta_t$, and $v_a + \delta_r + \delta_t$.

2.3.4 FIRST OVERTONES, CYCLIC MOLECULES

The first overtone stretching vibrations of methylene groups of strained-ring cyclic compounds such as cyclopropane occur near 6135 cm^{-1} (1630 nm). The effects of various substituents on the ring have been studied by several authors. Gassman and Zalar[21] list the band positions of 37 cyclopropane derivatives. Gassman[22] also published a table of first overtone CH band positions of aliphatic nortricyclene derivatives. These overtones were at slightly lower wavenumber maxima than the cyclopropanes — about 6024 cm^{-1} (1660 nm).

The C–H stretch first overtone region of cyclohexane has two strong peaks at 5697 cm^{-1} (1755 nm) and 5791 cm^{-1} (1727 nm). A number of smaller, additional peaks were also seen by curve resolution.[23] The two strongest peaks are probably $2v_a$ and $2v_s$[24] with the intensity of the symmetric band being intensified by Darling–Dennison resonance.

Cyclopentane, cyclobutane, cyclopropane, and many other cyclic molecules have been studied. In the fundamental region, the frequency increases as the ring becomes more strained, and the first overtones reflect this trend. Cyclopentane's two strongest first overtone peaks are at 5730 cm^{-1} (1745 nm) and 5834 cm^{-1} (1714 nm) for example, as compared to cyclohexane's 5697 and 5791 cm^{-1}.

Inductive effects on cyclopropane's first overtone have been extensively documented.[25,26] There is a shift from 6158 cm^{-1} (1624 nm) to 6060 cm^{-1} (1650 nm) in moving from strongly electron-withdrawing groups such as cyano to strongly electron-inducing groups such as methyls.

A terminal epoxy ring is one case in which methylene C–H groups are influenced by the strained structure and an oxygen atom. Such epoxides, such as epichlorohydrin and 1,2-diisobutylene oxide, have a sharp absorption band at about 6060 cm^{-1} (1650 nm) with a molar absorptivity of about 0.2 l/mol-cm.[27]

2.3.5 HIGHER-ORDER OVERTONES, CYCLIC MOLECULES

For cyclohexane, the second overtone of the C–H stretch is probably a combination of normal and local mode effects, with the two major bands being related to oscillators localized on the axial and equatorial C–H bonds.[28] The larger peak, near 8290 cm^1 for cyclohexane, 8434 cm^{-1} for cyclopentane, and 9116 cm^{-1} (1097 nm)[29] for cyclopropane, has been used for characterizing various cyclic structures from petroleum fractions.[30]

TABLE 2.5
Normal-Mode Band Assignments for Chloroform and Bromoform

	δ	ν	$\nu + \delta$	$\nu + 2\delta$	2ν	$2\nu + \delta$	$2\nu + 2\delta$	3ν	$3\nu + \delta$
CHCl$_3$	1215	3019	4215	5374	5910	7090	8229	8676	9836
CHBr$_3$	1143	3020	4145	5240	5907	7024	—	8677	9760

Higher overtones of cyclopentane, cyclobutane, and cyclopropane have also been studied with regard to their information content on equatorial and axial conformations.[31] It is generally acknowledged that these higher overtones are explained by local mode theory.

The second overtone of the C–H stretch connected to an epoxy ring has been used for quantitative work.[32] The peak appears at 8620 cm^{-1} (1160 nm).

2.3.6 COMBINATION BANDS, CYCLIC MOLECULES

The 4545-cm^{-1} (2220-nm) combination band of the cyclopropane group has been studied along with the first overtone, and has been found to also shift to higher frequencies with electron-withdrawing power.[33,34] The assignment is probably a $\nu + \delta$ combination.

Terminal epoxides have a similar combination band that has been used for quantitative analysis following the cure of epoxy resins.[35] It appears at about 4525 cm^{-1} (2210 nm) and has an absorptivity of about 1.5 l/mol-cm.[36]

2.4 METHINE GROUPS

The absorption of the single methine proton in the presence of methyl and methylenes is generally too small to be observed and is little mentioned in the literature. In the mid-infrared, the fundamental peak is near 2900 cm^{-1}, between the methyl and methylene doublets. This would suggest that its first overtone was also in the envelope of small peaks in the 5882–5555-cm^{-1} (1700–1800-nm) region.

Wheeler[37] gives the second overtone of an aliphatic methine to be 8163 cm^{-1} (1225 nm), and a second combination band to be at 6944 cm^{-1} (1440 nm).

The isolated aliphatic C–H group has been studied in halogenated compounds.[38] Table 2.5 lists the band assignments for chloroform and bromoform, expressed in terms of normal modes. The authors mention that some of these bands are split by the nondegeneracy of the deformation vibrations. These assignments are in agreement with those of Kaye.[39]

Iwamoto et al.[38] also discuss the effect of the isolation of the C–H group in the CHX$_3$ and CHX$_2$–CX$_2$–CHX$_2$ molecules, in comparison with the coupled modes of CHX$_2$–CHX$_2$ molecules. In the latter case, additional peak splitting is observed.

In general, the spectral features of vibrations of haloalkanes are affected by the mass of the halogen atom and the force constant of the carbon–halogen bond. If a halogen is bound to the same carbon atom as the hydrogen, the neighboring C–H overtone bands are shifted to a lower wavelength and their intensities are intensified.[40]

2.5 MODEL COMPOUND COMPARISONS

Figures 2.8 through 2.12 illustrate the general C–H absorption spectra–structure correlation by using three model compounds: trimethylpentane, *n*-decane, and toluene. Note that there are 12 methyl C–H bonds and 6 methylene C–H bonds in trimethyl-pentane; and there are 6 methyl C–H bonds and 16 methylene C–H bonds in *n*-decane. In the toluene molecule, there are 3 methyl C–H bonds; 0 methylene C–H bonds; and 5 aromatic C–H bonds.

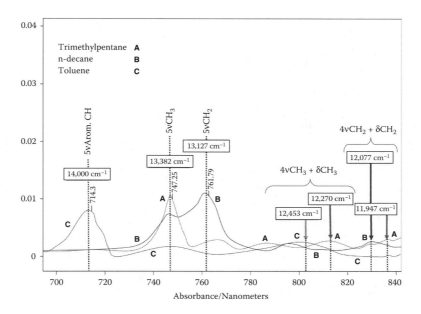

FIGURE 2.8 Specific illustration of the fourth overtone (5 ν) near-infrared C–H stretch as aromatic C–H, methyl C–H, and methylene C–H stretching band positions in nanometers (nm), with band positions labeled in wavenumbers (cm⁻¹). Within the same figure is shown the third overtone (4 ν) combination region. For the third overtone (4 ν) combination region, methyl stretching and methyl bending combination bands indicated between 800 and 820 nm (12,500 to 12,195 cm⁻¹) are observed. Also observed are methylene stretch and the bending combination bands between 830 and 840 nm (12,048 to 11,905 cm⁻¹).

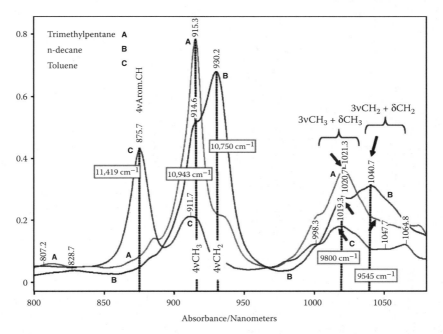

FIGURE 2.9 Demonstration of the third overtone (4 ν) harmonic and the second overtone (3 ν) combination region, illustrating the relative intensities and band positions. The third overtone (4 ν) near-infrared C–H stretching aromatic, methyl, and methylene band positions are shown. A region of second overtone (3 ν) combination bands is also seen in this figure.

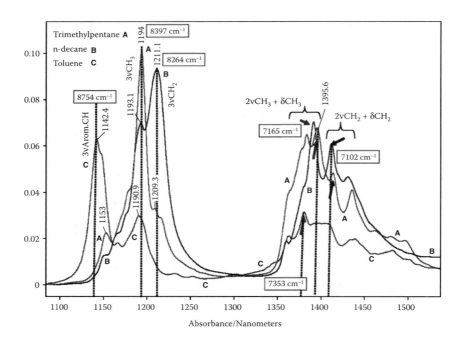

FIGURE 2.10 Illustration of the second overtone (3 ν) spectral region. Also observed are the first overtone (2 ν) stretching and bending combination bands for methyl and methylene as labeled in the figure.

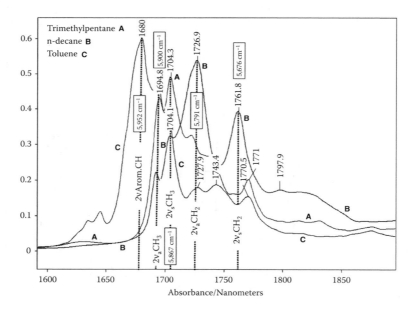

FIGURE 2.11 A spectrum of the first overtone (2 ν) spectral region. For these bands we are able to observe both asymmetric *and* symmetric C−H stretching bands, which were not visible in spectra from the second (3 ν), third (4 ν), or fourth (5 ν) overtone C−H stretching band regions (Figure 2.8 to Figure 2.10). Within this spectral region one observes the aromatic C−H as a single band, but clearly observes the asymmetric methyl and symmetric methyl stretching as separate bands. We also see that two bands are present for the asymmetric and symmetric methylene stretching.

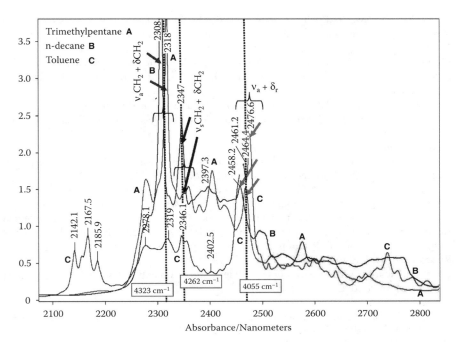

FIGURE 2.12 A spectrum of the first combination region for the model compounds. From these spectra of the model compounds, one observes the complexity of the combination region starting from about 2100 nm (i.e., 4762 cm^{-1}) and extending to near 3000 nm (i.e., 3333 cm^{-1}). There has not been comprehensive assignment work done for each of these bands. A few assignments are labeled in the figure. For example, the asymmetric methylene C–H stretching plus C–H bending combination band is shown near 2308 nm (i.e., 4333 cm^{-1}). This figure also shows the symmetric methylene C–H stretching plus C–H bending combination band near 2347 to 2350 nm (i.e., 4261 to 4255 cm^{-1}). This spectral region is also the location for the second overtone (2 v) methyl bending band near 2476 nm (i.e., 4039 cm^{-1}). Additional discussion of this spectral region was presented earlier in this chapter.

REFERENCES

1. Abney, R.E. and Festing, R.E., On the influence of the atomic grouping in the molecules of organic bodies on their absorption in the infrared region of the spectrum, *Phil. Trans. R. Soc. London*, 172, 887–918, 1881.
2. Ellis, J.W., The near-infrared absorption spectra of some organic liquids, *Phys. Rev.*, 23(1), 48–62, 1924.
3. Tosi, C. and Pinto, A., Near-infrared spectroscopy of hydrocarbon functional groups, *Spectrochim. Acta*, 28A, 585–597, 1972.
4. Luty, T. and Rohleder, J.W., The near-infrared quantitative spectroscopy of CH$_3$-groups in methyl derivatives of benzene, *Ann. Soc. Chim. Polonorum.*, 41, 975–983, 1967.
5. Bernard, U. and Berthold, P.H., Application of NIR spectrometry for the structural group analysis of hydrocarbon mixtures, *Jena Rev.*, 20(5), 248–251, 1975.
6. Rose, F.W., Quantitative analysis, with respect to the component structural groups of the infrared (1 to 2μ) molal absorptive indices of 55 hydrocarbons, *J. Res. Natl. Bur. Stand.*, 20, 129–157, 1938.
7. Evans, A., Hibbard, R.R., and Powell, A.S., Determination of carbon-hydrogen groups in high molecular weight hydrocarbons by near-infrared absorption, *Anal. Chem.*, 23, 1604–1610, 1951.
8. Salzer, R., Weise, D., and Boenisch, U., Strukturgruppenanalyse an kohlenwasserstoffgemischen im NIR, *Z. Chem.*, 25(7), 263–265, 1985.
9. Wheeler, O.H., Near-Infrared spectra of organic compounds, *Chem. Rev.*, 59, 629–666, 1959.

10. Fang, H.L. and Swofford, R.L., Highly excited vibrational states of molecules by thermal lensing spectroscopy and the local mode model. II. Normal, branched, and cyclo-alkanes, 73(6), 2607–2617, 1980.

11. Murray, I., in *The NIR Spectra of Homologous Series of Organic Compounds in Near- Infrared Diffuse Reflectance/transmittance Spectroscopy*, Hollo, J., Kaffka, K.J., and Gonczy, J.L. (Eds.), Akademiai Kiado, Budapest, 1987, pp. 13–28.

12. Murray, I. and Williams, P.C., Chemical principles of near-infrared technology in the agricultural and food industries, Williams, P. and Norris, K. (Eds.), American Association of Cereal Chemists, St. Paul, MN, 1987, pp. 17–34.

13. Murray, I., op. cit.

14. Ricard-Lespade, L., Longhi, G., and Abbate, S., The first overtone of C–H stretchings in polymethylene chains: a conformationally dependent spectrum, *Chem. Phys.*, 142, 245–259, 1990.

15. Buback, M. and Harfoush, A.A., Near-infrared absorption of pure n-heptane between 5000 cm^{-1} and 6500 cm^{-1} to high pressures and temperatures, *Z. Naturforsch.*, 38a, 528–532, 1983.

16. Wexler, A.S., Spectrometric determination of proton (CH) in organic compounds by integrated intensity measurements in the vCH second overtone region, *Anal. Chem.*, 40, 1868–1872, 1968.

17. Fang, H.L., and Swofford, R.L., op. cit.

18. Ricard-Lespade, L., op. cit.

19. Kaye, W., Near-infrared spectroscopy: a review, spectral identification and analytical applications, *Spectrochim. Acta*, 6, 257–287, 1954.

20. Murray, I., op. cit.

21. Gassman, P.G. and Zalar, F.V., Near-infrared studies. The dependence of the cylcopropyl C-H stretching frequency on inductive effects, *J. Org. Chem.*, 31, 166–171, 1966.

22. Gassman, P.G., Near-infrared studies: the cyclopropyl group, *Chem. Ind.*, 740–741, April 21, 1962.

23. Olinger, J.M., Ph.D. dissertation, University of California, Riverside, 1989.

24. Bonanno, A.S. and Griffiths, P.R.J., Short-wave near-infrared spectra of organic liquids, *Near-Infrared Spec.*, 1, 13–23, 1993.

25. Gassman, P.G. and Zalar, F.V., op.cit.

26. Weitkamp, H. and Korte, F.Z., Zur indentifizierung der cyclopropanstruktur, *Tetrahedron*, 20, 2125–2135, 1964.

27. Goddu, R.F. and Delker, D.A., Determination of terminal epoxides by near-infrared spectrophotometry, *Anal. Chem.*, 30, 2013–2016, 1958.

28. Fang, H.L. and Swofford, R.L., Photoacoustic spectroscopy of vibrational overtones in polyatomic molecules, *Appl. Optics.*, 21(1), 55–60, 1982.

29. Wheeler, O.H., op. cit. p. 639.

30. Evans, A. et al., op. cit.

31. Wong, J.S., MacPhail, R.A., Moore, C.B., and Strauss, H.L., Local mode spectra of inequivalent C-H oscillators in cycloalkanes and cycloalkenes, *J. Phys. Chem.*, 86, 1478–1484, 1982.

32. Dannenberg, H., Determination of functional groups in epoxy resins by near-infrared spectroscopy, *SPE Trans.*, 78–88, January 1963.

33. Weitkamp, H., op. cit.

34. Simmons, H.E., Blanchard, E.P., and Hartzler, H.D., The infrared spectra of some cyclopropanes, *JOC*, 295–301, 1966.

35. Weyer, L.G., Near-infrared spectroscopy of organic substances, *Appl. Spectrosc. Rev.*, 21(1,2), 1–43, 1985.

36. Goddu, R.K., op. cit.

37. Wheeler, O.H., op. cit. p. 638.

38. Iwamoto, R., Nara, A., and Matsuda, T., Near-infrared combination and overtone bands of CH in CHX_3, CHX_2-CHX_2 and CHX_2-CX_2-CHX_2, *Appl. Spectrosc.*, 59(11), 1393–1398, 2005.

39. Kaye, W., op. cit. pp. 266–267.

40. Wheeler, O.H., op. cit.

3 Alkenes and Alkynes

3.1 LINEAR ALKENES, OVERTONES

The first overtone of the C–H stretch next to a double bond occurs at a higher wavenumber (lower wavelength) than saturated C–H stretch absorptions. This peak is strong and distinct in some structures, particularly the methylene group of terminal double. In most cases, however, it is weak and difficult to locate especially in the presence of methyl groups. The band position is near 6100–6200 cm^{-1} (1640–1612 nm).

The C–H stretch first overtone of terminal methylene groups of vinyl and vinylidene structures is isolated enough that it can be used in traditional quantitative analysis. Figure 3.1 provides one example, and Table 3.1 provides some typical peak locations. Goddu[1] provides tables of absorptivities for the first overtone absorption of the terminal methylene group in a variety of compounds and solvents. Molar absorptivities are about 0.2–0.5 l/mol-cm. Put another way, a 100-ppm amount of methylene gives an absorbance of 0.01 in a 10-cm cell. Analyses using this peak to measure the vinyl content of acrylate monomers,[2] butadienes,[3] and edible oils[4] have been reported.

In studies of wavelength displacement of deuterium substitution, hexadeuteropropylene ($CD_3CD=CD_2$) gives an asymmetric stretching of the CD_2 at 4610 cm^{-1} (2170 nm).[5]

In acrylates, in addition to the first overtone near 6000 cm^{-1} (1665 nm), there is a band at 6167 cm^{-1} (1620 nm), with overtones at 9020 cm^{-1} (1109 nm) and 11,800 cm^{-1} (847 nm), used by Gerasimov and Snavely[6] to study diacrylate and dimethacrylate reactions. Although this is said to be a combination band, it is a combination of two C–H stretch vibrations, rather than a C–H with another type of vibration.

The C–H stretch first overtone of a *cis* double bond has also been found to be distinct. It appears near 5963 cm^{-1} (1677 nm) and is isolated from the vinyl overtones.[8] Figure 3.2 illustrates the features of *cis* C–H absorptions in comparison to a *trans* configuration.

The second overtone of the terminal methylene C–H stretch can also be used for quantitative analysis. It occurs at about 9260 cm^{-1} (1080 nm). In 1-alkenes, it is 8897–8944 cm^{-1} (1118–1124 nm), with a molar absorptivity of 0.004–0.018 l/mol-cm. The peak is at 9017 cm^{-1} (1109 nm) in allyl stearate and a doublet at 8787–9009 cm^{-1} (1110–1138 nm) in alkyl acrylates. The peak is at 9091 cm^{-1} (1100 nm) in alkyl vinyl ethers and about 9059 cm^{-1} (1104 nm) in vinyl esters.[9]

The third overtone of terminal methylenes' C–H stretch is found at 11,390 cm^{-1} (878 nm) in 1-alkenes.[10] Its absorptivity is 0.002 l/mol-cm. The alkyl acrylate doublet occurs at 11,905 and 12,500 cm^{-1} (800 and 840 nm). The peak is at 10,776–11,360 cm^{-1} (880–928 nm) in alkyl vinyl ethers, and 10,788–10,929 cm^{-1} (915–927 nm) in vinyl ethers and vinyl esters. The fourth through seventh overtones of terminal methylene C–H stretch peaks have been studied using photoacoustic spectroscopy.[11] These were given as approximately 11,450, 14,090, 16,630, and 18,800 cm^{-1}, respectively (873, 710, 600, and 532 nm, respectively). In the spectrum of propylene, this peak is split into a doublet of two peaks that are *cis* and *trans* to the methyl group.

3.2 LINEAR ALKENES, COMBINATIONS/BENDING MODES

A distinctive set of bands in the combination region exists for vinyl double bonds. As seen in Figure 3.1, there are three peaks at about 4482 cm^{-1} (2230 nm), near 4600 cm^{-1} (2170 nm), and 4670–4780 cm^{-1} (2090–2140 nm).

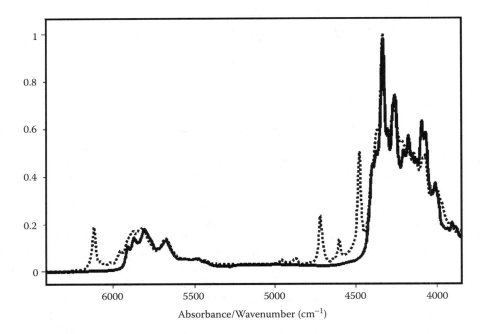

FIGURE 3.1 Hexane (solid) and 1-hexene (dotted).

The vinyl group in polybutadienes has been studied more extensively than simple alkenes, and it has a similar spectrum with peaks at about 4484, 4597, and 4660 cm^{-1}. Bands at 4717 and 4481 cm^{-1} in polybutadiene have been assigned as second overtones of the symmetric and asymmetric bending vibrations of the CH_2 of the uncoupled vinyl group, as interpreted by a local mode model.[12] There are no corresponding fundamental vibrations for these peaks at 1469 and 1572 cm^{-1} though;

TABLE 3.1
Alkene Functional Groups

Functional Group	CH Peak in cm^{-1}	CH Peak in nm
Vinyl (hexene) $CH_2=CH-$	6120	1635
Vinylidene (3-chloro-2-methyl-1-propene) $CH_2=C<$	6130	1631
Vinyl on N (1-ethenyl 2-pyrrolidinone)	6170	1621
Vinyl on O (1-ethenyloxybutane)	6200	1613
Vinyl and vinylidene (2-methyl-1,3-butadiene)	6130/6140	1631/1629
Acrylate $R-O-\overset{\|}{\underset{O}{C}}-CH=CH_2$	6169[a]	1621[a]
Methyne (1-hexyne) $CH\equiv$	6536	1530

[a] According to Gerasimov and Snavely,[7] this peak is actually a combination of vibrations, one quantum of energy from each of the two C–H bonds of the vinyl group, and the actual first overtone is at 5998 cm^{-1} (1667 nm).

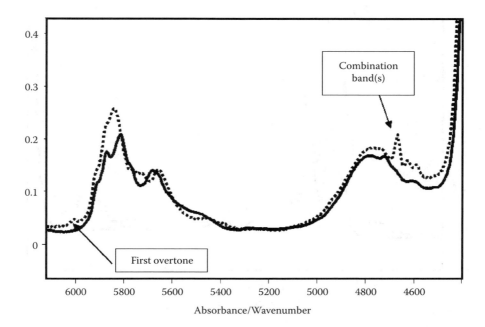

FIGURE 3.2 *Cis* and *trans* compounds, *cis*-3-hexane-1-ol (dotted curve) and *trans*-2-hexene-1-ol (solid). Note: Peaks specific to *cis* double bonds are indicated with arrows.

only the overtones are observed. The overtones at 2X and all other even multiples are not observable because they coincide with stronger C–H-stretching vibrations. The vinyl peak, at 4600 cm^{-1}, has also been attributed to a bending mode by Snavely and Angevine, as has the 4717 cm^{-1} peak.

These vinyl bands have also been discussed by McManis and Gast.[13] The 4482-cm^{-1}/2230-nm peak has been used to quantify the vinyl group having an absorptivity of about 0.87 l/mol-cm. The peak shifts slightly in allyl esters and ethers and splits into two peaks when a carbonyl is present. The 4600-cm^{-1}/2170-nm peak was observed in all vinyl compounds except vinyl esters. Its absorptivity is about 0.2 l/mol-cm. The 4700-cm^{-1}/2100-nm peak has an absorptivity of about 0.45 l/mol-cm.

Higher-order overtones of the 4481- and 4717-cm^{-1} peaks have also been noted at 7346 cm^{-1} (1360 nm) and 7750 cm^{-1} (1290 nm). The 4600-cm^{-1} peak also shows a second member of a progression at 7508 cm^{-1} (1332 nm). Its absorptivity has been reported to be about 0.01 l/g-cm.

As in the C–H stretch overtones, *trans* double bonds do not have distinctive absorptions in the combination region, but *cis* double bonds do. There is a strong *cis* peak near 4673 cm^{-1} (2140 nm) that can be used for quantification, and a second peak near 4587 cm^{-1} (2180 nm). These peaks are pointed out in Figure 3.2.

3.3 CYCLIC DOUBLE BONDS

When cyclohexene is compared to cyclohexane, the most readily apparent difference occurs at the combination band near 4670 cm^{-1} (2140 nm), similar to the *cis* double bond in straight-chain alkenes. There are additional bands in the overtone area as well. A small doublet near 6050 cm^{-1} (1650 nm) might also be attributed directly to the C–H stretch next to the double bond. See Figure 3.3.

Cyclopentene also has a distinct combination band near 4660 cm^{-1} (2145 nm) and four peaks near 6000 cm^{-1} (1667 nm) that are not present in cyclopentane. At 8777 cm^{-1} (1140 nm) there is

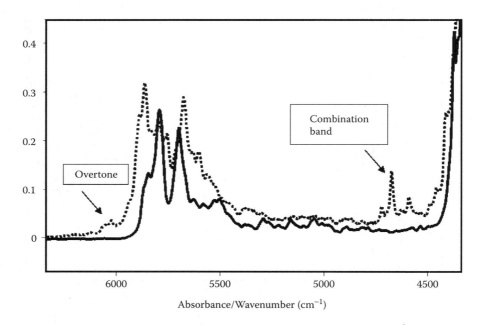

FIGURE 3.3 Cyclohexane (solid curve) and cyclohexene (dotted).

a single, very distinctive C–H second overtone due to the C–H of the double bond.[14] Wong and associates have analyzed and discussed the fifth overtone spectra of cyclic alkenes.[15]

The first overtone of the C–H stretch of the double bond in norbornenes was determined to be in the range of 5970–6080 cm^{-1} (1645–1675 nm).[16] The attribution of this absorption band to the C–H groups next to the double bonds and not the bridgehead C–H was confirmed by examination of model compounds that lacked the vinyl C–Hs and retained the bridgehead C–H. The effect of a number of different substituents on the band position was also studied.

3.4 DIENES

Dienes are described by the following structure and comprise a set of molecules with somewhat unique properties.

As in aromatic compounds, dienes demonstrate multiple bands associated with conjugated double-bond and C–H stretching. The mid-infrared spectra of conjugated polyenes exhibit a band in the 1000- to 900-cm^{-1} (10,000- to 11,111-nm) region, which does not have overtones or combination bands observed in the near-infrared. This band is indicative of *cis* and *trans* groups within the different polyenes.[17] The near-infrared bands associated with ν H–C=C are observed near 6110 cm^{-1} (1637 nm), 5960 cm^{-1} (1678 nm), 4710 cm^{-1} (2123 nm), 4595 cm^{-1} (2176 nm), and 4470 cm^{-1} (2237 nm) and are very similar to those shown for 1-hexene in Figure 3.1. The 3ν C=C band is observed near 4950 cm^{-1} (2020 nm). Note that the Appendix material provides more examples of model spectra as well as tables showing precise band positions.

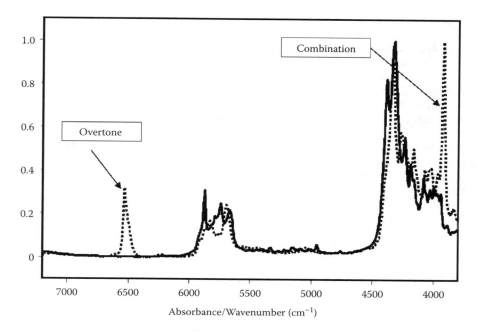

FIGURE 3.4 1-Hexyne (dotted curve), and 2-hexyne (solid).

3.5 ALKYNES AND ALLENES

The C–H stretch of a proton next to a triple bond occurs near 6500 cm^{-1} (1538 nm) and is very distinctive and isolated from other C–H absorption peaks. There is also a combination band at 3930 cm^{-1} (2545 nm) that could be the sum of the C–H stretch and the C–H wag. These are shown in Figure 3.4. 2-Hexyne, which has no proton on its triple bond, is shown for comparison. A second overtone of the C–H stretch occurs at about 9600 cm^{-1} (1042 nm).

Allene, $H_2C=C=CH_2$, is said to have its first C–H overtone at 6139 cm^{-1} (1629 nm) and a combination band at 6031 cm^{-1} (1658 nm). Second and third overtones have been reported at 8620 cm^{-1} and 11,765 cm^{-1}, respectively (1160 and 850 nm, respectively).[18]

REFERENCES

1. Goddu, R.F., Near-infrared spectrophotometry, in *Advances in Analytical Chemistry and Instrumentation*, Vol. 1, Reilly, C.N. (Ed.), Interscience, New York, 1960, pp. 347–424.
2. Gerasimov, T.G. and Snavely, D.L., Vibrational overtone spectroscopy of ethylene glycol diacrylate and ethylene glycol dimethacrylate, monomer and polymer, *Appl. Spectrosc.*, 56(2), 212–216, 2002.
3. Snavely, D.L. and Angevine, C., Near-infrared spectrum of polybutadiene, *J. Polym. Sci.: Part A*, 34, 1669–1673, 1996.
4. Murray, I., The NIR spectra of homologous series of organic compounds in near- infrared diffuse reflectance/transmittance spectroscopy, Hollo, J., Kaffka, K.J., and Gonczy, J.L. (Eds.), Akademiai Kiado, Budapest, 1987, pp. 13–28.
5. Kaye, W., Near-infrared spectroscopy: a review, Spectral identification and analytical applications, *Spectrochim. Acta*, 6, 257–287, 1954.
6. Gerasimov, T.G. and Snavely, D.L., op. cit.
7. Gerasimov, T.G. and Snavely, D.L., ibid.
8. McManis, G.E. and Gast, L.E., Near-infrared spectra of long chain vinyl derivatives, *JAOCS*, 48, 310–313, 1971.

9. McManis, G.E. and Gast, L.E., op. cit.

10. Wheeler, O.H., Near-Infrared spectra of organic compounds, *Chem. Rev.*, 59, 629–666, 1959.

11. Fang, H.L. and Swofford, R.L., Photoacoustic spectroscopy of vibrational overtones in polyatomic molecules, *Appl. Opt.*, 21(1), 55–60, 1982.

12. Snavely, D.L. and Angevine, C., op. cit.

13. McManis, G.E. and Gast, L.E., op. cit.

14. Buback, M. and Voegele, H.P., *FT-NIR Atlas*, VCH, New York, 1993.

15. Wong, J.S. and Moore, C.B., Inequivalent C-H oscillators of gaseous alkanes and alkenes in laser photoacoustic overtone spectroscopy, *J. Chem. Phys.*, 77(2), 603–615, 1982.

16. Gassman, P.G. and Hooker, W.M., Near-infrared studies. Norbornenes and related compounds, *JACS*, 87(5), 1079–1083, 1965.

17. Colthup, N.B., The interpretation of the infrared spectra of conjugated polyenes in the 1000–900 cm^{-1} region, *Appl. Spectrosc.*, 25(3), 368–371, 1971.

18. Wheeler, O.H., op. cit., p. 642.

4 Aromatic Compounds

4.1 BENZENE

The near-infrared (IR) spectrum of benzene has been thoroughly described by Kaye[1] and others. Benzene is such a highly symmetrical molecule that only a few of its vibrations are infrared active. However, the inactive modes participate in the overtones and combinations of the near-infrared. An explanation for the nomenclature of the fundamental vibrational modes used by Kaye is provided in Figure 4.1 from Avram and Mateescu.[2] Only the vibrations labeled "IR" have fundamental IR bands. The band assignments for the first combination and first overtone regions are provided in Table 4.1, using the Herzberg nomenclature as described in Figure 4.1.

The first overtone doublet near 6000 cm^{-1} (1670 nm) is actually composed primarily of IR-inactive C–H-stretching vibrations. As described in Table 4.1 and Figure 4.1, both of the bands near 6000 cm^{-1} are composed of a Raman active vibration and a vibration that is neither IR nor Raman active. The benzene NIR spectrum is shown in Figure 4.2.

The second overtone peak at 8834 cm^{-1} (1132 nm) represents the second overtone of the IR-active C–H stretch (vibration 12 in Herzberg notation and Figure 4.1).[3] A nearby peak at 8770 cm^{-1} (1140 nm) has been assigned as a combination of twice the C–H stretch plus either 14 and 19, or 8 and 9. A third overtone is at 11,442 cm^{-1} (874 nm) neat, and 11,364 cm^{-1} (880 nm) in carbon tetrachloride. The fourth overtone is at about 14,000 cm^{-1} (714 nm).

The strong peak at 4050 cm^{-1} (2469 nm) is therefore a combination of vibrations 14 and 15 of Figure 4.1, a C–H stretch and C–H bending as described in the figure. The series of absorptions near 4660 cm^{-1} (2146 nm) combine C–C stretching with C–H stretching. Combination peaks near 3640 cm^{-1} (2747 nm) are C–H stretching combined with C–C or C–H-wagging vibrations.

Deuterated benzene (C_6D_6) has bands at 1900, 1560, 1375, and 1010 nm, which correspond to the first, second, third, and fourth overtones of C-D stretching, respectively.[4]

4.2 SUBSTITUTED AROMATICS — ALKYL

Substitution on the benzene ring reduces the symmetry of the molecule, potentially allowing for additional vibrational peaks. In the case of alkyl aromatics, there is also the addition of the alkyl absorptions. Figure 4.2 compares the near-infrared spectra of benzene and toluene. The primary differences appear to be the addition of the methyl combination bands near 4300 cm^{-1} (2300 nm) and the methyl first overtones near 5800 cm^{-1} (1750 nm). There is also a noticeable peak at 3836 cm^{-1} (2607 nm) that could be related to the sum of a C–H stretch and one of the ring-wagging vibrations involving a ring with five adjacent protons.

Alkyl groups on the benzene ring displace the CH overtone band to a higher wavenumber (longer wavelength) due to their electropositive nature. Also, as seen in Figure 4.4, the strong peak in the region of 4060 cm^{-1} (2460 nm) is a doublet in mono-, di-, and tri-substituted methylbenzenes. The lower wavenumber peak in this doublet has been attributed to the C–H stretch/CC bending mode described in the preceding text, whereas the peak near 4080 cm^{-1} (2452 nm) has been assigned as the second overtone of the symmetric methyl-bending vibration.[5] A linear relationship was found between the number of methyl groups substituted on the benzene ring and the intensity of this characteristic absorption band. The 4080-cm^{-1} band gave a good measure for aromatic compounds with 1 to 4 methyl groups. In higher-substituted methyl-benzenes such as durene and pentamethylbenzene, the first overtone band at 5660 cm^{-1} (1767 nm) provides better predictions.

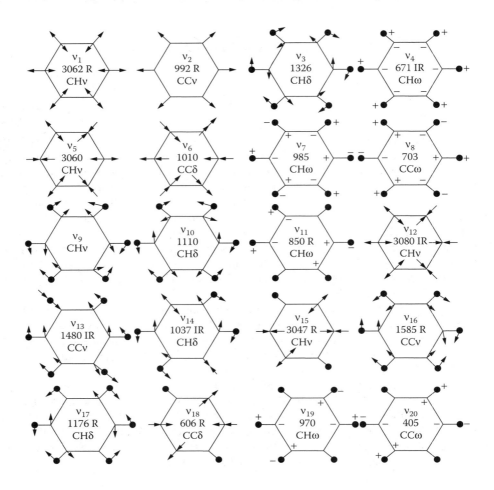

FIGURE 4.1 Herzberg symbols.

With regard to higher vibrational states, the second overtone of the aromatic C–H stretch in alkylated benzenes occurs at 8734 cm^{-1} (1145 nm), whereas the alkyl C–H stretch occurs at 8389 cm^{-1} (1192 nm). Approximately 5- to 10-nm shifts were observed to lower frequency (longer wavelength) as compared with benzene's second CH overtone band.

4.3 SUBSTITUTED AROMATICS — NONALKYL

As shown in Figure 4.5, electronegative groups such as halogens or nitro groups produce shifts to higher wavenumber (lower wavelength), whereas electropositive substituents (alkyl groups) generate a displacement to lower wavenumber.

In an investigation of NIR absorptions using photoacoustic experiments, the tentative assignments of 11 substituted benzenes have been reported.[6] These included three benzonitriles, five acetonitriles, and three benzylbromides. An example of a typical group of band assignments, for 2,2,4-trichloroacetophenone, given in this work is:

4057 cm^{-1} (2465 nm)   sum of 2927 and 1070 cm^{-1}
4141 cm^{-1} (2415 nm)   sum of 3100 and 1070 cm^{-1}

TABLE 4.1
Benzene Band Assignments

Assignment	Vibrations	Calculated Wavenumber	Observed Wavenumber	Observed Wavelength (nm)
18 + 5	CCδ+CHν	3666	3613 (3613)	2768
12 + 18	CHν+CCδ	3705	3640 (3643)	2747
8 + 5	CCω+CHν	3731	3693 (3697)	2708
4 + 1	CHω+CHν	3733	Masked	—
7 + 5	CCω+CHν	3763	3722 (3735)	2687
11 + 12	CCω+CHν	3948	3937 (3935)	2540
19 + 15	CCω+CHν	4017	3960 (3958)	2525
15 + 6	CHν+CCδ	4047	3980 (3986)	2513
14 + 15	CCδ+CHν	4084	4050 (4060)	2469
12 + 2	CHν+CCν	4091	Shoulder	—
14 + 1	CCδ+CHν	4099	Shoulder	—
15 + 10	CHν+CHδ	4157	4155 (4175)	2407
17 + 5	CCδ+CHν	4238	4190 (4198)	2387
12 + 17	CHν+CHδ	4277	4252 (4263)	2352
12 + 3	CHν+CHδ	4425	4360 (4379)	2294
13 + 15	CCν+CHν	4532	4532 (4549)	2206
13 + 1	CCν+CHν	4547	4570 (4584)	2188
16 + 5	CCν+CHν	4656	4615 (4625)	2167
12 + 16	CHν+CCν	4695	4642 (4644)	2154
15 + 9	CHν+CCν	4695	4655 (4675)	2148
15 + 5	CHν+CHν	6107	5920 (5914)	1689
12 + 1	CHν+CHν	6161	5985 (5988)	1671

Note: The wavenumbers in parentheses represent peak positions determined by the author in Reference 1.

4246 cm⁻¹ (2355 nm)	sum of 3100 and 1162 cm⁻¹
4320 cm⁻¹ (2315 nm)	sum of 2950 and 1406 cm⁻¹
4494 cm⁻¹ (2225 nm)	sum of 2950 and 1479 cm⁻¹
4651 cm⁻¹ (2150 nm)	sum of 3100 and 1590 cm⁻¹
4938 cm⁻¹ (2025 nm)	sum of 2998/2950 and 2 × 991 cm⁻¹
5141 cm⁻¹ (1945 nm)	3 × 1716 cm⁻¹
5291 cm⁻¹ (1890 nm)	sum of 3100 and 2 × 1070 cm⁻¹
5540 cm⁻¹ (1805 nm)	sum of 2950 and 2 × 1282 cm⁻¹
5731 cm⁻¹ (1745 nm)	2 × 2858 cm⁻¹
5935 cm⁻¹ (1685 nm)	2 × 2950 cm⁻¹
6173 cm⁻¹ (1620 nm)	2 × 3100 cm⁻¹
6431 cm⁻¹ (1555 nm)	sum of 2998 and 2 × 1708 cm⁻¹
6667 cm⁻¹ (1500 nm)	2 × 2950 plus 727 cm⁻¹
6897 cm⁻¹ (1450 nm)	2 × 2950 plus 991 cm⁻¹
7273 cm⁻¹ (1375 nm)	2 × 2998 plus 1282 cm⁻¹
7752 cm⁻¹ (1290 nm)	2 × 2998 plus 1708 cm⁻¹
8230 cm⁻¹ (1215 nm)	3100 plus 3 × 1708 cm⁻¹
8969 cm⁻¹ (1115 nm)	3 × 2950/2998 cm⁻¹

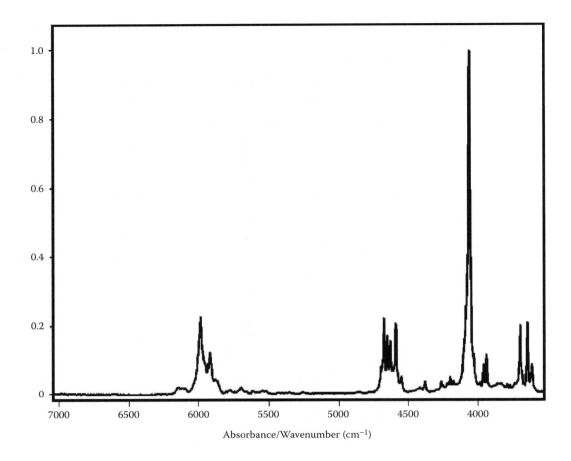

FIGURE 4.2 Spectrum of benzene.

4.4 OTHER AROMATIC COMPOUNDS

The presence of a heteroatom in an aromatic ring splits the C–H overtone peaks rather than shifting them. For example, in liquid pyridine there is a doublet at each overtone with the lower energy peak attributed to the C–H-stretching absorptions of the 2 and 6 positions, and the higher-energy peak due to the 3, 4, and 5 positions.[7] This is shown in Figure 4.6. These peaks appear at 5835 and 5867 cm^{-1} (1714 and 1704 nm) for the 2,6-positions, and 5907 and 5956 cm^{-1} (1693 and 1679 nm) for 3, 4, and 5. Bini et al. suggest that the 3-, 4-, and 5-position peaks appear as only one peak due to overlap.[8]

The highest wavenumber peak in the region near 6000 cm^{-1} has been assigned by Bini et al. as a combination of two C–H-stretching vibrations, both involving the 2,6-positions. The smaller peaks at 6060, 6105, and 6139 cm^{-1} are also combinations, of 2,6 and 3,5; 3,5 and 3,5; and 3,5 and 3,5 stretching vibrations.

Figure 4.6 also shows the first combination spectral region of benzene and pyridine near 4500–4700 cm^{-1}. The distinctive set of four large peaks and several smaller ones are considerably shifted for pyridine, relative to benzene. The assignments for these peaks are given in Table 4.1. All are combinations involving a C–H stretch and another vibration.

Thiophene, furan, and pyrrole also have two peaks at each overtone.[9] The spectra of gases of these compounds all shift to higher energy (lower wavelength) relative to the spectra of liquids.

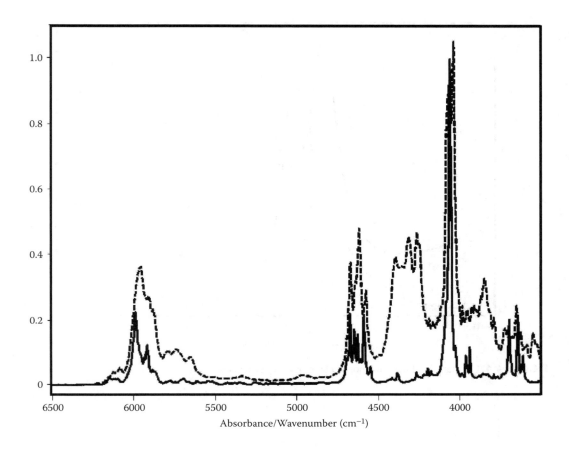

FIGURE 4.3 Comparison of benzene (solid curve) and toluene (dotted curve).

The second overtones of C–H stretch peaks for liquid pyridine occur at 8785 cm^{-1} (1138 nm) and 8651 cm^{-1} (1156 nm). The third and fourth overtones are at 11,296 and 11,474 cm^{-1} (885 and 872 nm) and 13,810 and 14,037 cm^{-1} (724 and 712 nm), respectively. The lower frequency band has been assigned to C–H stretches at the 2 and 6 positions, and the higher frequency band was attributed to the C–H stretches at the 3, 4, and 5 positions.

In a study of methyl-substituted pyridines, the aryl regions of the overtones show a simplified structure having one peak progression for each nonequivalent C–H. The methyl regions of the methylpyridines show complex profiles. The band profile in 3- and 4-methylpyridine is similar to that of toluene because the methyl groups of these compounds are free rotors, and all have a low-energy barrier to rotation. However, the methyl band profiles of 2-methylpyridine are complex, and these patterns indicate that vibration-torsional coupling is an important contributor to the complex structure.[10]

The near-infrared spectrum of the five-membered ring aromatic organometallic compound ferrocene has also been examined.[11] The C–H stretch first overtone of the mid-infrared 3080-cm^{-1} peak occurs at 6105 cm^{-1} (1638 nm). A combination peak at 4495 cm^{-1} (2225 nm) has been assigned as a combination of the 1400-cm^{-1} C–C stretch mid-infrared peak and the C–H stretch. A doublet centered at 4167 cm^{-1} (2400 nm) appears to be due to the combination of the asymmetric and symmetric ring-breathing vibrations (1000 and 1100 cm^{-1} in the mid-infrared) with C–H stretching. Further, a strong peak at 3945 cm^{-1} (2535 nm) is a combination of the 810-cm^{-1} C–H bending with the 3080-cm^{-1} C–H stretch. Lewis also suggested that two smaller peaks observed at

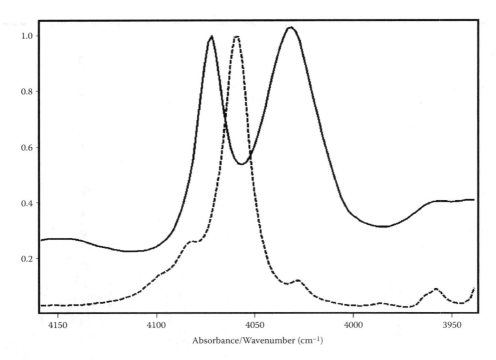

FIGURE 4.4 Comparison of benzene (dotted curve) and *m*-xylene (solid curve) in the 4100-cm^{-1} region.

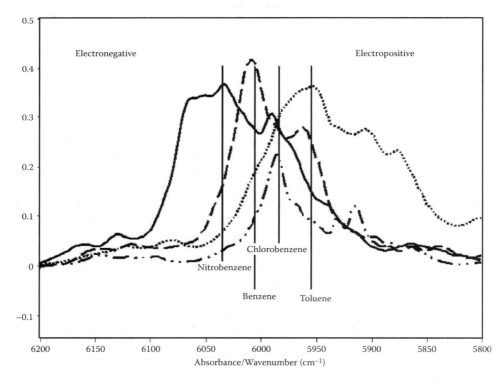

FIGURE 4.5 First overtone peak shifting with electronegativity of substituents on benzene ring. (From Weyer, L.G. and Lo, S.-C., in *Handbook of Vibrational Spectroscopy*, Chalmers, J.C. and Griffiths, P.R., John Wiley & Sons, 2002. With permission.)

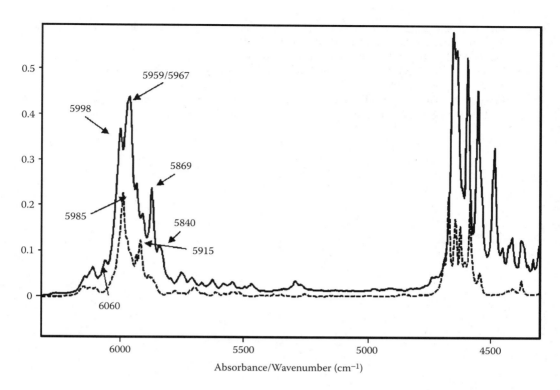

FIGURE 4.6 Comparison of benzene (dashed) and pyridine (solid) spectra.

4367 cm^{-1} (2290 nm) and a broad peak near 4800 cm^{-1} (2083 nm) are combinations of the C–H stretch with two inactive IR bands at 1257 cm^{-1} (C–H bend) and 1560 cm^{-1} (C–C stretch), respectively.

Heterocyclic aromatics with a prominent N–H band include pyrrole and indole where the nitrogen is attached to a single hydrogen. For these compounds, the strong first overtone N–H is apparent at 6835 cm^{-1} (1463 nm), and the combination band is observed at 4715 cm^{-1} (2121 nm). The N–H overtone can be shifted by as many as 400 cm^{-1} when other substituents are present, such as in 2,2-dimethyl-thiazolidine, where the band is observed near 6450 cm^{-1} (1550 nm). When N–H is present in the heterocyclic compound, a second-overtone bending band may be observed as a very weak shoulder near the 4715-cm^{-1} (2121-nm) position. Other differences between heterocyclics are found in the C–H region between 6300 cm^{-1} and 5700 cm^{-1} (1587 nm to 1754 nm).

REFERENCES

1. Kaye, W., Near-infrared spectroscopy, *Spectrochim. Acta*, 6, 257–287, 1954.
2. Avram, M. and Mateescu, G.D., *Infrared Spectroscopy: Applications in Organic Chemistry*, Wiley-Interscience, New York, 1970, p. 202.
3. Bassi, D., Menegotti, L., Oss, S., Scotini, M., and Iachello, F., The 0 to 3 C–H stretch overtone of benzene, *Chem. Phys. Lett.*, 207(2,3), 167–172, 1993.
4. Kaye, W., op. cit.
5. Luty, T. and Rohleder, J.W., Near-infrared quantitative spectroscopy, *Rocz. Chemi.*, 41, 975–983, 1967.
6. Sarma, T.V.K., Sastry, C.V.R., and Santhamma, C., The photoacoustic spectra of substituted benzenes in the near-infrared region, *Spectrochim. Acta*, 43A, 1059–1065, 1987.
7. Snavely, D.L., Overly, J.A., and Walters, V.A., Vibrational overtone spectroscopy of pyridine and related compounds, *Chem. Phys.*, 201, 567–574, 1995.

8. Bini, R., Foggi, P., and Della Valle, R.G., Vibrational analysis of C–H stretching overtones in pyridine and 2,6-lutidine, *J. Phys. Chem.*, 95, 3027–3031, 1991.

9. Snavely, D.L., Overly, J.A., and Walters, V.A., op. cit.

10. Proos, R.J. and Henry, B.R., Overtone investigation of methyl-substituted pyridines, *J. Phys. Chem. A*, 103, 8762–8771, 1999.

11. Lewis, L.N., The analysis of the near-infrared of some organic and organo-metallic compounds by photoacoustic spectroscopy, *Organomet. Chem.*, 234, 355–365, 1982.

5 Hydroxyl-Containing Compounds

5.1 O–H FUNCTIONAL GROUPS

The hydroxyl group has been extensively studied in the near-infrared region, particularly with respect to hydrogen bonding. The advantages of longer pathlength cells for liquids and direct analysis of solids have contributed to the utility of near-infrared for hydrogen-bonding studies in a great variety of mixtures and environments. Also, the near-infrared offers a special advantage over the mid-infrared in the measurement of mixtures of water and other hydroxyl-containing compounds because the combination peaks of water and other hydroxyls are well separated from each other in the near-infrared.

5.2 ALCOHOLS

5.2.1 First Overtone Region

The first overtone of a free hydroxyl group in dilute CCl_4 solution or a low-density gas is at about 7090 cm^{-1} (1410 nm). This peak is at different positions for primary, secondary, and tertiary alcohols, as seen in Figure 5.1. Primary and secondary butanols can be split into doublets by rotational isomerization.[1] The splits are better seen in Figure 5.2, in the second derivation spectra of the same spectral region. Maeda et al. observed an additional peak in the first overtone region when they subtracted the spectrum at a lower temperature from that at a higher one. They felt that temperature effects further separated species that were weakly bonded to the carbon tetrachloride solvent and a terminal free OH of a self-associated species.

A hydroxyl first overtone peak may also be split due to interactions with the electrons of phenyl rings or halogens, as in benzyl alcohol, for example.

In hydrogen-bonded alcohols, there is a broad peak in the 1460–1600 nm region (6850 cm^{-1}–6240 cm^{-1}), which has been generally attributed to the first overtone of the hydroxyl. This bonded OH peak appears to be broader as a first overtone than its fundamental, as was observed in a series of spectra taken at different temperatures.[2] On the other hand, the nonbonded OH first overtone is stronger relative to the bonded species than it is in the fundamental, and can be readily detected even when not visible in the fundamental region.[3] Figure 5.3 illustrates the differences between the bonded and nonbonded hydroxyls of methanol. The broad bonded OH is indistinct in the neat spectrum, whereas the nonbonded OH in a dilute carbon tetrachloride solution is very sharp and strong.

Temperature- and concentration-dependent variations in self-association of 1-octanol have also been studied by two-dimensional (2-D) Fourier transform near-infrared correlation spectroscopy.[4] The population of the free OH groups increases with temperature, reportedly reaching 13% at 80°C. The molar absorptivities of the first and second overtones of the monomer were found to be similar in several non-hydrogen-bonding solvents such as carbon tetrachloride, heptane, and octane. The first overtone's absorptivity was about 1.7 l/mol-cm, which agrees with Goddu's data.[5] The absorptivity of the second overtone is about one twentieth that of the first.

Hydrogen bonding of alcohols occurs to different extents with solvents with varying bonding strengths. The spectra of methanol in a series of solvents from carbon tetrachloride through pyridine

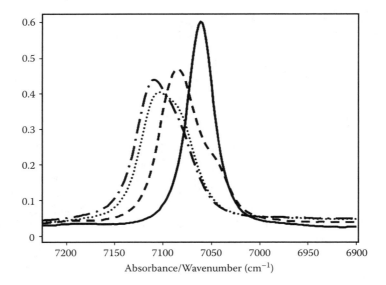

FIGURE 5.1 First overtone of OH peak in butanols. The dot-dashed curve is isobutanol (a primary alcohol), the dotted curve is 1-butanol, the dashed curve is *sec*-butanol, and the solid curve is *tert*-butanol. (From Weyer, L.G. and Lo, S.-C., in *Handbook of Vibrational Spectroscopy*, Chalmers, J.C. and Griffiths, P.R., John Wiley & Sons, 2002. With permission.)

have been studied.[6,7] The first overtone region exhibits two major bonded OH peaks near 6800 and 7100 cm^{-1} (14,700 and 1280 nm) when an alcohol is in a solvent capable of hydrogen bonding. Bell and Barrow have shown that the ratio of these two peaks is not affected by temperature or dilution (except for the formation of dimer at very high concentrations). They also show that the second peak is not due to an OH/CH combination band. They suggest that the two peaks are both monomeric hydrogen-bonded-alcohol first overtones, and that the vibrational energy level has been split into two, creating a double minimum in the potential energy curve.

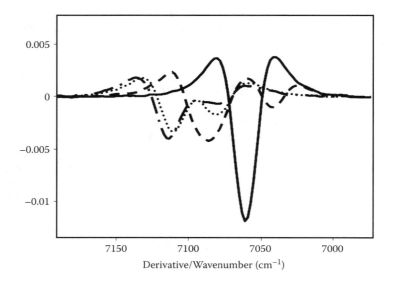

FIGURE 5.2 Derivative spectra of butanols.

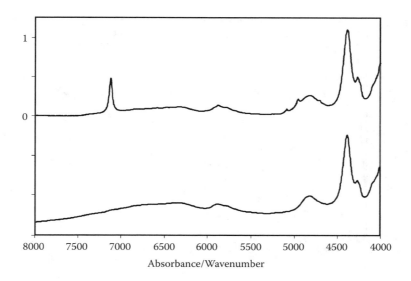

FIGURE 5.3 Methanol in CCl$_4$ (top) and neat (bottom) spectra.

5.2.2 HIGHER-ORDER OVERTONES

The second overtone of the nonbonded OH stretch occurs at about 10,400 cm^{-1} (960 nm), and the third at about 13,500 cm^{-1} (740 nm) for simple alcohols. The second overtone has also been used for a number of hydrogen-bonding studies.[8] Variations in the structure of the alcohol result in splitting of the band and systematic shifts. Second overtones of the OH stretch appear to have less interference from CH combination bands than first overtones, and can therefore be more useful for thermodynamic studies.[9] Additional overtones of the nonbonded hydroxyl stretch of alcohols, using gaseous ethanol as a model, are the fourth at 16,700 cm^{-1} (600 nm) and the fifth at 19,500 cm^{-1} (510 nm).[10] Additional bonded hydroxyl bands include the OH-stretch second overtone at 9550 cm^{-1} (1047 nm), a combination of the first overtone of the OH stretch and twice the methyl CH deformation at 9386 cm^{-1} (1065 nm), and a combination of the OH-stretch first overtone plus three times the CO stretch at 9720 cm^{-1} (1029 nm).[11]

5.2.3 COMBINATION BANDS

A distinctive combination region of alcohols occurs between 4550 and 5550 cm^{-1} (1800–2200 nm). This broad peak, a combination of OH stretching and OH bending, has been used in a large number of quantitative analyses including hydroxyl number of polymers, alcohols in the presence of water, and ethylene vinyl alcohol content in copolymers. As shown in Figure 5.3, it is present to some extent in both dilute solution and neat spectra of hydroxyl-containing compounds.

In the dilute solution spectrum of methanol, three sharp peaks are observed superimposed on the broader continuum. These can be seen clearly in Figure 5.3. The sharp peaks have been assigned to OH stretch plus CH bending at 5090 cm^{-1} (1965 nm), OH stretch plus OH bending at 4960 cm^{-1} (2017 nm), and OH stretch plus CO stretch at 4710 cm^{-1} (2124 nm).[12] There is also a second combination of OH-stretching and a different OH-bending mode at about 3970 cm^{-1} (2520 nm). These sharp peaks are at different positions in other alcohols. In the neat spectrum, these peaks are seen as only one broad band centered at about 4770 cm^{-1} (2100 nm). In the mid-infrared, a diffuse OH association band related to deformation occurs at about 1420 cm^{-1}, which can account for the 4770-cm^{-1} near-infrared band (1420 cm^{-1} + 3350 cm^{-1} = 4770 cm^{-1}). The diffuse mid-infrared band is said to disappear in dilute solutions of alcohols, where hydrogen

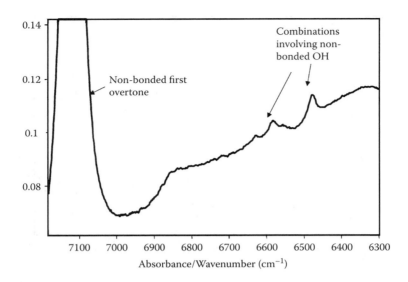

FIGURE 5.4 OH combination peaks near the much larger first overtone peak.

bonding does not occur.[13] However, the broad combination peak does not disappear completely in the near-infrared.

There has been some confusion over whether combinations of OH stretch and C–H stretch in the 6330-cm^{-1} (1580-nm) region contribute to the broad hydrogen-bonded OH overtone peak.[14,15] Although Kaye appears to assign the 6330-cm^{-1} peak to this combination, he also says that combination bands involving C–H stretch and OH stretch are very weak because stretching vibrations involving different atoms do not couple well unless their frequencies are nearly the same or the groups involve double bonds or rings. Kaye's apparent assignment may be simply a function of how one reads his chart, as the two small OH/CH peaks that are observable in his dilute CCl_4 spectrum overlap the broad, bonded OH-stretch first overtone. Figure 5.4 shows the small peaks near the much larger first overtone peak. Bell and Barrow[16] show that the small peaks (one in ethanol and two in methanol, due to coupling with asymmetric and symmetric C–H stretch) are not present in CD_3OH dissolved in carbon tetrachloride. Czarnecki et al.[17] suggest that the series of peaks between 6240 cm^{-1} (1600 nm) and 7100 cm^{-1} (1410 nm) are all due to OH-stretch first overtones of different aggregates, such as monomers, dimers, and polymers of the alcohols, but Czarnecki also says in a later article that a band at 6500 is a CH-stretch and nonbonded OH-stretch combination.[18] Czarnecki et al. found that the 6330-cm^{-1} peak decreases with temperature when the monomeric 7100-cm^{-1} peak increases, which supports their theory that it is due primarily to OH overtones. There are apparently contributions from both hydroxyl overtones and CH/OH combinations involved in the broad envelope, although the bonded OH first overtone probably predominates. The evidence of Davies and Rutland does not contradict this statement.

There has also been a suggestion that a 6319-cm^{-1} (1580-nm) broad peak in ethylene-vinyl alcohol copolymers is due to a combination of OH stretching and 2 OH bending.[19] However, this peak is generally considered another hydrogen-bonded OH first overtone.

Higher-wavenumber combination bands include one that combines the first overtone of the OH stretch and twice the methyl CH deformation at 9386 cm^{-1} (1065 nm), and a combination of the OH stretch first overtone plus three times the CO stretch at 9720 cm^{-1} (1029 nm).[20]

5.3 PHENOLS

5.3.1 OVERTONES

Phenols show absorptions in the same regions as aliphatic alcohols, with first overtones of the O–H near 6940–7140 cm^{-1} (1400–1440 nm). As seen in Figure 5.5 the first overtone of monomeric phenol in CCl$_4$ is a single peak at 7040 cm^{-1} (1420 nm). Ortho-halogens can show a doublet due to *cis/trans* isomerism where the *cis* form is stabilized by internal hydrogen bonding between the hydroxyl and the halogen.[21,22] The *cis* form, represented by the large peak shown in Figure 5.5, is shifted to about 6830 cm^{-1} (1464 nm). The smaller *trans* peak is barely discernable in the figure. According to Wheeler, the ratio of the *cis* and *trans* peaks at different temperatures provides a measure of the stabilizing energy of the *cis* form.

The second overtone of dilute phenol in carbon tetrachloride is at 10,000 cm^{-1} (1000 nm), and the third at 13,250 cm^{-1} (750 nm). The second and third overtones of the *o*-halogenated phenols are doublets due to *cis/trans* configurations. For *o*-iodophenol, for example, the second overtone peaks are at 9910 and 1029 cm^{-1} (1009 and 972 nm, respectively), and the third overtones at 12,788 and 13,400 cm^{-1} (782 and 746 nm, respectively).

Due to the difference in acidity of phenol relative to simple aliphatic alcohols, its spectra in different solvents are quite different from those of methanol and ethanol.[23] In solvents of low hydrogen-bonding capability, phenol shows both free and bonded OH first overtones because the solvent cannot bond the phenol completely. In solvents that are more capable of hydrogen bonding, phenol behaves like the alcohols, except that the sequence is accelerated; for example, the spectrum of phenol in *N*,*N*-dimethylformamide appears similar to the spectrum of ethanol in pyridine.

5.3.2 PHENOLS COMBINATIONS

In a study of combination bands of various halogenated phenols, Wulf et al.[24] found that most of the combinations involved the OH and bending or twisting modes of the aromatic ring, or the OH group itself. This conclusion was based on the observation that pentachlorophenol, which had no CH groups, showed very similar patterns to the other model species. Rospenk et al.[25] compared the spectra of phenol and phenol-OD to show that a series of strong combination bands between

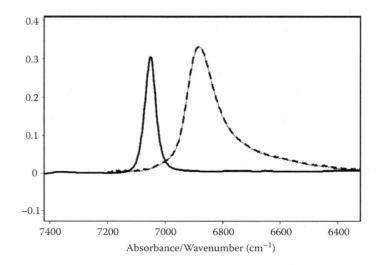

FIGURE 5.5 First overtone of phenol in CCl$_4$ (solid curve) and neat orthochlorophenol (dashed).

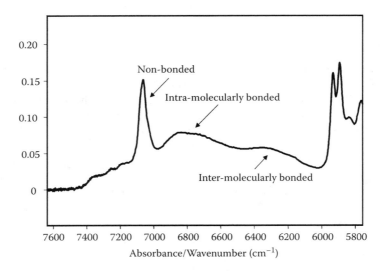

FIGURE 5.6 Spectrum of 2-methyl-2,4-pentanediol showing how glycols and other compounds having more than one hydroxyl have the opportunity for different types of hydrogen bonding.

5210 cm^{-1} (1920 nm) and 4760 cm^{-1} (2100 nm) all involve the hydroxyl group but no CH. The 5210-cm^{-1} and 5080-cm^{-1} peaks were assigned to OH stretch plus interactions with the ring, the 4949 cm^{-1} to OH stretch plus OH deformation, and the 4867 cm^{-1} and 4783 cm^{-1} were probably due to OH stretch plus C-O stretches with some interactions with other vibrations.[26]

5.4 MULTIPLE HYDROXYL COMPOUNDS

Glycols and other compounds having more than one hydroxyl have the opportunity for different types of hydrogen bonding. These include many natural compounds such as sucrose, starch, and cellulose. Even in dilute solution, a diol will show an intramolecularly hydrogen-bonded hydroxyl peak.[27] Both bonded and nonbonded peaks are observed. The spectrum of 2-methyl-2,4-pentanediol in Figure 5.6 shows a nonbonded peak at 7065 cm^{-1} (1415 nm) and an intramolecularly bonded peak at about 6850 (1460 nm). A small 6370-cm^{-1} (1570-nm) intermolecularly bonded peak also appears. Based on information from the fundamental region, it is likely that the spacing between the nonbonded and the intramolecularly bonded peaks increases with the number of carbon atoms between the two hydroxyls.[28,29]

Carbohydrates in general may have a free OH-stretch absorption near 6940 cm^{-1} (1440 nm). This band has been reported in crystalline sucrose, for example, and has been assigned specifically to the C$_4$ hydroxyl within a crystalline matrix.[30] Trott et al.[31] discuss four different OH first overtone bands in carbohydrates in different solvent systems, using a monomer (glucose) and its polymer (glycogen) as models.

5.5 HYDROPEROXIDES

The first overtone of the nonbonded hydroxyl peak of a hydroperoxide in dilute solution is far enough removed from that of acids and alcohols that it can be used for quantitative analysis.[32] The peak is at about 6850 cm^{-1} (1460 nm) as compared to 7100 cm^{-1} (1410 nm) for alcohols. The spectrum of cumene hydroperoxide shown in Figure 5.7 shows a splitting of the hydroxyl

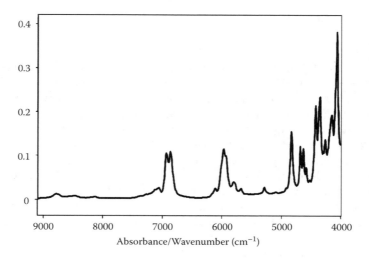

FIGURE 5.7 Cumene hydroperoxide.

peak due to interaction with the benzene ring. This doublet changes to a singlet in methylene chloride. There is also a combination band that probably involves the OH stretch and OH bending at about 4850 cm[-1] (2060 nm), again in dilute solution. These two bands were found to be characteristic of hydrogen peroxides derived from oxidized fatty esters in dilute carbon tetra-chloride solutions.[33]

Normally, H_2O_2 is available as a dilute solution in water, and its OH first overtone is hydrogen bonded and obscured by that of water. However, the combination band near 4850 cm[-1] (2060 nm) can be clearly observed, as shown in Figure 5.8.

High overtones of the OH stretch, such as the fifth at 619 nm and the sixth at 532 nm, have been used in studying photoactivated dissociation of t-butyl hydroperoxide.[34] Excitation at these wavelengths creates a vibrationally hot electronic ground state.

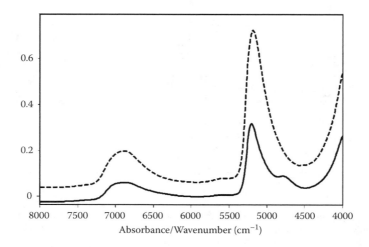

FIGURE 5.8 Hydrogen peroxide solution (solid curve) compared to water (dashed curve).

5.6 OH IN CARBOXYLIC ACIDS

5.6.1 OVERTONES

The OH group associated with monomeric carboxylic acid has a sharp nonbonded or free-stretching first overtone at about 6920 cm^{-1} (1445 nm). This band has been observed in the high-temperature (92°C) spectrum of octanoic acid.[35] As seen in the inset of Figure 5.9, there is some overlap of this band with CH combination bands. This overlap appears more pronounced in acids than in alcohols because the acid hydroxyl absorptivity is significantly lower than the alcohol absorptivity. As the temperature is lowered or the solution becomes more concentrated, the formation of dimers with hydrogen-bonded hydroxyl groups broadens the OH stretch overtone and moves it to a lower frequency (longer wavelength) as it does in the mid-infrared. This is illustrated in Figure 5.9, a temperature series. Although one can usually observe the broad dimeric hydroxyl band in the mid-infrared, in the near-infrared it appears only to shift the baseline upward in the region from about 1700 nm to at least 2200 nm. Spectral subtraction reveals a very broad and shallow curve in this region.

The second overtone of the nonbonded carboxylic acid hydroxyl is at about 10,000 cm^{-1} (1000 nm), and the third at about 12,500 cm^{-1} (800 nm).[36]

5.6.2 OH IN CARBOXYLIC ACIDS, COMBINATION BANDS

A peak near 5290 cm^{-1} (1890 nm) due to OH stretch combined with C=O stretch is readily observable in the high-temperature curves of Figure 5.9. Also, a doublet at about 4630 cm^{-1} (2160 nm) and 4695 cm^{-1} (2130 nm) can be seen. This is due to a C–H stretch and C=O stretch combination, and has been split because of a rotational mode.

Other, smaller features in the spectrum of monomeric carboxylic acids observed in the gaseous state[37] include small peaks near 8200 cm^{-1} (1220 nm) due to a combination between the first overtone of the OH stretch and a COH bending mode, 8070 cm^{-1} (1240 nm) due to the first overtone of the OH stretch and the CO stretch, 7600 cm^{-1} (1315 nm) due to the first overtone of the OH stretch and OCO bending, and 6500 cm^{-1} (1540 nm) due to OH stretch and C–H stretch.

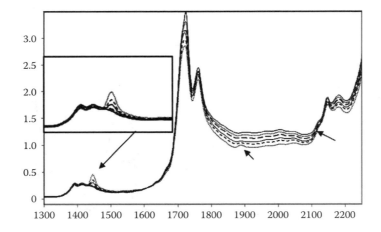

FIGURE 5.9 Carboxylic acid hydroxyl peaks. Arrows near 1890 and 2130/2160 point out combination bands relating to the nonbonded hydroxyl. (From Weyer, L.G. and Lo, S.-C., in *Handbook of Vibrational Spectroscopy*, Chalmers, J.C. and Griffiths, P.R., John Wiley & Sons, 2002. With permission.)

Additional small peaks include those at about 4950 cm^{-1} (2020 nm), a combination of the OH stretch and CH bending; 4800 cm^{-1} (2080 nm), the combination of OH stretch and COH bending; 4710 cm^{-1} (2120 nm), the combination of C–H stretch and C=O stretch; 4680 cm^{-1} (2140 nm), the combination of OH stretch and CO stretch; 4210 cm^{-1} (2380 nm), the combination of OH stretch and OCO bending; 4184 cm^{-1} (2390 nm), the combination of C–H stretch and COH bending; and 4050 cm^{-1} (2470 nm), the combination of C–H stretch and CO stretch. All of these assignments are for formic acid in the gaseous state. Acids with more complex hydrocarbon structures would have more combination bands, and the region from 4000–4500 cm^{-1} (2200 nm to 2500 nm) can become quite complex.

Spectra of dimeric carboxylic acids have a different set of small peaks due to the shifts in the OH and C=O structures. In the gaseous spectrum of formic acid, small peaks appear at about 5630 cm^{-1} (1780 nm) due to a combination of C–H stretch, CH bending, and CO stretch; 4700 cm^{-1} (2130 nm) due to a combination of C=O stretch and C–H stretch; 4500 cm^{-1} (2220 nm) due to OH stretch and OH–O bending; 4460 cm^{-1} (2240 nm) due to OH stretch and CH bending; 4340 cm^{-1} (2300 nm) due to C–H stretch and OH–O bending; 4310 cm^{-1} (2320 nm) due to OH stretch and CO stretch; and 4160 cm^{-1} (2400 nm) due to C–H stretch and COH bending.

5.7 SILANOLS

The first overtone of the OH stretch of a silanol group in fused silica optical fibers is at about 7200 cm^{-1} (1390 nm).[38] See Figure 5.10 for a comparison with water. This band has been resolved into four peaks, two major and two minor, representing three different OH stretch peaks in different environments and one combination with a 280-cm^{-1} SiO$_2$ fundamental that is Raman active. The second overtone shows a similar pattern centered at about 10,600 cm^{-1} (940 nm), and the third near 13,800 cm^{-1} (725 nm).

Combination band assignments given by Stone and Walrafen[38] for bands at 4100 cm^{-1} (2440 nm), 4450 cm^{-1} (2250 nm), and 4520 cm^{-1} (2210 nm) involve an OH stretch with one of the SiO$_2$ fundamentals. Combinations involving twice the frequency of an OH stretch with the SiO stretch near 800 cm^{-1} were noted at 7920 cm^{-1} (1260 nm) and 8065 cm^{-1} (1240 nm), and one band at three times an OH stretch plus the 800-cm^{-1} band was observed at 11,500 cm^{-1} (850 nm).

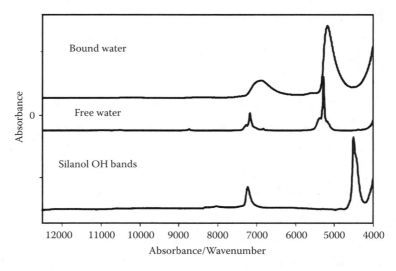

FIGURE 5.10 Comparison of silanol bands in silica with those of water.

5.8 ENOLS AND ENOLATES

The enols contain both unsaturated C=C double bonds as well as an associated O−H group. These alkene-alcohol and associated anions have the general chemical descriptor as

$$:\ddot{O}H$$
$$-C \diagdown\diagdown_{C-}$$

An enolate is an enol-based anion formed by the presence of an enol in a basic medium. An enolate anion, readily formed in the presence of base, is as

$$:\ddot{O}H \qquad :\ddot{O}:^-$$
$$-C \diagdown\diagdown_{C-} \quad \rightleftharpoons \quad -C \diagdown\diagdown_{C-}$$

Important near-infrared bands associated with enols as phenols are the same as those given for phenols.[39] In Figure 5.11, two example enols as phenols are overlaid from 7200 cm⁻¹ to 3800 cm⁻¹ (1389 nm to 2632 nm). The measured bands associated with phenols include 3ν OH as a broad band from 10,200 to 9700 cm⁻¹; 2ν OH as a broad band from 7200 to 6000 cm⁻¹; and the ν OH + δ OH combination envelope located at 5000 to 4500 cm⁻¹. Specific bands associated with enols are very weak and are mainly associated with C=C-H from the aromatic ring. These can be found in the various chapters and spectra demonstrating aromatic and alkene C-H associated -C=C-H bands.

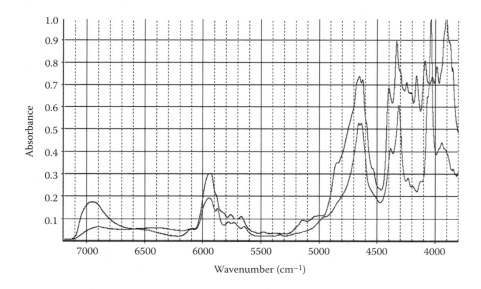

FIGURE 5.11 Two examples of enols as phenols. The top spectrum is 2-ethyl phenol, and the bottom is 3-methyl phenol. (Spectra used by permission from *NIR Spectra of Organic Compounds*, Wiley-VCH. ISBN 3-527-31630-2.)

REFERENCES

1. Czarnecki, M., Maeda, H., Ozaki, Y., Susuki, M., and Iwahasi, M., A near-infrared study of hydrogen bonds in alcohols — comparison of chemometrics and spectroscopic analysis, *Chemometry Intelligent Lab. Sys.*, 45, 121–130, 1999.

2. England, L., Schloeberg, D., and Luck, W.A.P., The meaning of the different O-H bandshapes in infrared fundamental and overtone regions of H-bonded methanol, *J. Mol. Struct.*, 143, 325–328, 1986.

3. Boueron, C., Peron, J.J., and Sandorfy, C., A low-temperature study of sterically hindered associated alcohols, *J. Phys. Chem.*, 76(6) 864–868, 1972.

4. Czarnecki, M.A., Effect of temperature and concentration on self-association of octan-1-ol studied by two-dimensional Fourier Transform near-infrared correlation spectroscopy, *J. Phys. Chem.*, 104(27), 6356–6361, 2000.

5. Goddu, R.F., Near-infrared spectrophotometry, in *Advances in Analytical Chemistry and Instrumentation,* Vol. 1, Reilly, C.N. (Ed.), Interscience, New York, 1960, pp. 347–424.

6. Bell, C.L. and Barrow, G.M., Hydrogen bond: an experimental verification of the double minimum potential, *J. Chem. Phys.*, 31(2), 300–307, 1959.

7. England-Kretzer, L., Fritzche, M., and Luck, W.A.P., The intensity change of IR OH bands by H-bonds, *J. Mol. Struct.*, 175, 277–282, 1988.

8. Wheeler, O.H., Near-infrared spectra of organic compounds, *Chem. Rev.*, 59, 629–666, 1959.

9. England-Kretzer, L. et al., op. cit.

10. Fang, H.L. and Swofford, R.L., Molecular confirmers in gas-phase ethanol: a temperature study of vibrational overtones, *Chem. Phys. Lett.*, 105(1), 5–11, 1984.

11. Rai, S.B. and Srivatava, P.K., Overtone spectroscopy of butanol, *Spectrochim. Acta,* Part A, 55, 2793–2800, 1999.

12. Kaye, W., Near-infrared spectroscopy, *Spectrochim. Acta*, 6, 257–287, 1954.

13. Colthup, N.B., Daly, L.H., and Wiberley, S.E., *Introduction to Infrared and Raman Spectroscopy,* Academic Press, New York, 1990.

14. Kaye, W., op. cit.

15. Davies, A.M.C. and Rutland, S.G., Identification of an OH, CH combination band in the near-infrared spectrum of ethanol, *Spectrochim. Acta*, 44A(11), 1143–1145, 1988.

16. Bell, C.L. and Barrow, G.M., op. cit.

17. Czarnecki, M.A., Maeda, H., Ozaki, Y., Susuki, M., and Iwahashi, M., Resolution enhancement and band assignments for the first overtone of OH stretching mode of butanols by two-dimensional near-infrared correlation spectroscopy. Part I: sec-butanol, *Appl. Spectrosc.*, 52(7), 994–1000, 1998.

18. Czarnecki, M.A., op. cit.

19. Iwamoto, R., Amiya, S., Saito, Y., and Samura, H., FT-NIR spectroscopic study of OH groups in ethylene-vinyl alcohol copolymer, *Appl. Spectrosc.*, 55(7), 864–870, 2001.

20. Rai, S.B. and Srivatava, P.K., Overtone spectroscopy of butanol, *Spectrochim. Acta*, Part A, 55, 2793–2800, 1999.

21. Wulf, O.R., Jones, E.J., and Deming, L.S., Combination frequencies associated with the first and second overtones and fundamental OH absorption in phenol and its halogen derivatives, *J. Chem. Phys.*, 8, 753–765, 1940.

22. Rospenk, M., Leroux, N., and Zeegers-Huyskens, Assignment of the vibrations in the near-infrared spectra of phenol-OH(OD) derivatives and application to the phenol-pyrazine complex, *J. Mol. Spectrosc.*, 183, 245–249, 1997.

23. Bell, C.L. and Barrow, G.M., op. cit.

24. Wulf, O.R., Jones, E.J., and Deming, L.S., op. cit.

25. Rospenk, M. et al., op. cit.

26. Rai, S.B. and Srivastava, P.K., op. cit.

27. Luck, W.A.P. and Ditter, W., Approximate methods for determining the structure of H_2O and HOD using near-infrared spectroscopy, *J. Phys. Chem.*, 74(21), 3687–3695, 1970.

28. Goddu, R.F., op. cit.

29. Luck, W.A.P. and Ditter, W., op. cit.

30. Davies, A.M.C. and Miller, C.E., Tentative assignment of the 1440-nm absorption band in the near-infrared spectrum of crystalline sucrose, *Appl. Spectrosc.*, 42(4), 703–704, 1988.

31. Trott, G.F., Woodside, E.E., Taylor, K.G., and Deck, J.C., Physicochemical characterization of carbohydrate-solvent interactions by near-infrared spectroscopy, *Carbohydr. Res.*, 27, 415–435, 1973.

32. Goddu, R.F., op. cit.

33. Holman, R.T., Nickell, C., and Privett, O.S., Detection and measurement of hydroperoxides by near-infrared spectrophotometry, *J. Am. Oil Chem. Soc.*, 35, 422–425, 1958.

34. Chandler, D.W. and Miller, J.A., A theoretical analysis of photoactivated unimolecular dissociation: the overtone dissociation of t-butyl hydroperoxide, *J. Chem. Phys.*, 81(1), 455–463, 1984.

35. Ozaki, Y., Liu, Y., Czarnecki, M.A., and Noda, I., FT-NIR spectroscopy of some long-chain fatty acids and alcohols, *Macromol. Symp.*, 94, 51–59, 1995.

36. Workman, J.J., Interpretive spectroscopy for near-infrared, *Appl. Spectrosc. Revs.*, 31(3), 251–320, 1996.

37. Morita, S. and Nagakura, S., Near-infrared spectra of hydrogen-bonded cyclic dimmers of some carboxylic acids, *J. Mol. Spectrosc.*, 41, 54–68, 1972.

38. Stone, J. and Walrafen, G.E., Overtone vibrations of OH groups in fused silica optical fibers, *J. Chem. Phys.*, 76(4), 1712–1722, 1982.

39. Goddu, R.F., Determination of phenolic hydroxyl by near-infrared spectrophotometry, *Anal. Chem.*, 30(12), 2009–2013, 1958.

6 Water

6.1 WATER

The analysis of water in various media and under various conditions has been a major part of the field of near-infrared (IR) spectroscopy since its inception. The strength of the near-infrared absorption bands, the unique water combination band at 1940 nm, and the sensitivity of the absorption bands to the environment of the water molecules have all contributed to the success of near-infrared to study and measure water.

6.2 LIQUID WATER, ICE, AND WATER VAPOR

A summary of the absorption bands of water in the 800–2500-nm region is provided in Table 6.1.[1] The vapor spectral details shown in the table were probably taken from a low-resolution spectrum, as a much larger number of water-vapor peaks have been reported by others.[2] As shown in the table, the accepted assignments for the two bands at about 6900 cm^{-1} (1450 nm) and 10,300 cm^{-1} (970 nm) indicate that these are combination bands involving the symmetric and asymmetric stretching modes of the water molecule. This observation has been made by comparison with high-resolution vapor spectra, which show that the combination bands are stronger than the overtones of either the symmetric or asymmetric stretch.[3] This assignment is further supported by consideration of the symmetry group of the vibrations.[4] It has also been suggested that the first overtone of the asymmetric stretch is accidentally degenerate with the sum of asymmetric and symmetric stretches in dilute solutions.[5] Therefore, although these two bands are generally referred to as the first and second overtones of the OH stretch, they are actually combination bands.

The strong 5150-cm^{-1} (1940-nm) peak is a combination of the asymmetric stretch and bending of the water molecule. In liquid water, there also are weak, broad combination bands near 5620 cm^{-1} (1780 nm) and 8310 cm^{-1} (1200 nm), and a second set near 11,800 and 13,000 cm^{-1} (840 and 740 nm, respectively).

The water peaks with maxima near 10,300 cm^{-1} (970 nm)[7], 8330 cm^{-1} (1200 nm),[8,9] and 6900 cm^{-1} (1450 nm) at room temperature shift towards higher wavenumber (lower wavelength) with increasing temperature, and appear to consist of an unresolved pair of peaks with an isosbestic point between them. See Figure 6.1 for an illustration of the temperature effect at the "first overtone." The isosbestic point for the 10,300-cm^{-1} (970-nm) band is 10,100 cm^{-1} (990 nm), for the 8330-cm^{-1} (1200-nm) band is about 8400 cm^{-1} (1190 nm),[10] and for the 6900-cm^{-1} (1450-nm) band is at about 6960 cm^{-1} (1440 nm).[11] The presence of an isosbestic point indicates that NIR calibrations for water independent of temperature may be possible by judicious wavelength choice.

An isosbestic point has also been studied for pressure changes of water at constant temperature.[12] A shift to higher wavenumber (lower wavelength) occurred with increasing pressure. The pressure isosbestic point for the 10,300-cm^{-1} peak was at about 10,200 cm^{-1} (980 nm).

There has been some controversy over whether liquid water contains water molecules in three states of hydrogen bonding: free; bonded through one, two, or three hydrogen bonds; or a continuum of different bond strengths or somewhere in between.[13] In general, temperature studies and data

TABLE 6.1
Absorption Peaks of Water, Ice, and Vapor

Ice (nm)	Ice (cm⁻¹)	Liquid Near Freezing Point (nm)	Liquid Near Freezing Point (cm⁻¹)	Liquid Near Boiling Point (nm)	Liquid Near Boiling Point (cm⁻¹)	Vapor (nm)	Vapor (cm⁻¹)	Assignment
800	12,500	770	13,000	740	13,500	723	13,831	$3\,v_1 + v_3$
909	11,000	847	11,800	840	11,900	823	12,151	$2\,v_1 + v_2 + v_3$
1025	9760	979	10,210	967	10,340	942	10,613	$2\,v_1 + v_3$
1250	7990	1200	8310	1160	8640	1135	8807	$v_1 + v_2 + v_3$
1492	6700	1453	6880	1425	7020	1380	7252	$v_1 + v_3$
1780	5620	1780	5620	1786	5600	—	—	$v_2 + v_3 + v_L$
1988	5030	1938	5160	1916	5220	1875	5332	$v_2 + v_3$

Note: For this table, v_1 is the symmetric stretch of the water molecule, v_2 is the bending mode, and v_3 is the asymmetric stretch. v_L was not defined in the original reference. It has been postulated to be an "intermolecular mode."[6]

treatments such as second derivatives and spectral resolution have indicated that there are separate peaks present for different states of hydrogen-bonded water molecules.[14] As pointed out by Maeda et al., the intensities and width of these peaks are affected by the anharmonicity of each, so that the molecules bonded to more adjacent molecules, having less anharmonicity, will be broader and weaker.

Figure 6.2 and Figure 6.3 show the spectra of liquid water and water in a "free" state, in carbon tetrachloride.

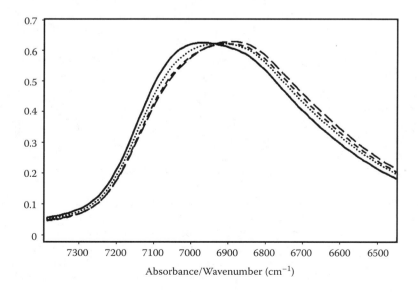

FIGURE 6.1 Water spectra at four temperatures from 25 to 65°C.

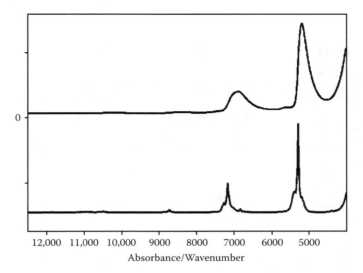

FIGURE 6.2 Free and bonded water spectra. Liquid water is shown in the top spectrum and water in carbon tetrachloride in the bottom spectrum.

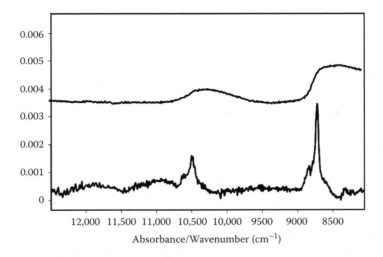

FIGURE 6.3 Water spectra, higher-frequency region. Liquid water is shown in the top spectrum and water in carbon tetrachloride in the bottom spectrum.

6.3 DEUTERIUM OXIDE (D₂O)

The near-infrared spectrum of fully deuterated water is very similar to that of H_2O, except that the peaks are all shifted because of the mass difference of the hydrogen atom. Table 6.2 summarizes the absorptions.[15] Figure 6.4 shows a portion of the near-infrared spectral region, illustrating the isotope shift of the $v_1 + v_3$ combination band (indicated by stars).

HDO is not of the same symmetry group as H_2O and D_2O, and therefore its peak shifts are not strictly isotope shifts. Also, liquid HDO is in equilibrium with H_2O and D_2O, and the contributions

TABLE 6.2
Band Assignments for H_2O, D_2O, and HDO

Liquid H_2O Near Freezing Point (nm)	Liquid H_2O Near Freezing Point (cm^{-1})	Liquid D_2O (nm)	Liquid D_2O (cm^{-1})	Liquid HDO (nm)	Liquid HDO (cm^{-1})	Assignment
770	13,000	—	—	—	—	$3 v_1 + v_3$
847	11,800	1190	8400	1000	10,000	$2 v_1 + v_2 + v_3$
979	10,210	1340	7470	1240	8065	$2 v_1 + v_3$
1200	8310	1620	6165	—	—	$v_1 + v_2 + v_3$
1453	6880	1970	5080	1670	5975	$v_1 + v_3$
1780	5620	2440	4100	—	—	$v_2 + v_3 + v_L$
1938	5160	2600	3830	2020	4945	$v_2 + v_3$

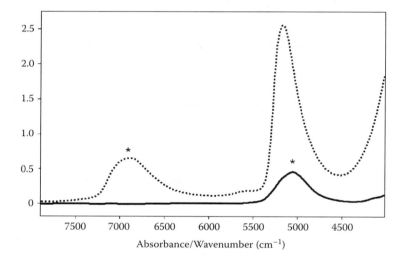

FIGURE 6.4 Comparison of water and deuterium oxide. The dotted curve is water, and the solid curve is deuterium oxide. Stars indicate the $v_1 + v_3$ combination band in both.

from these molecules must be subtracted in order to observe the spectrum of HDO. Note that the large combination band used most often for moisture measurement is out of the normal NIR range for deuterium oxide.

6.4 WATER IN VARIOUS SOLVENTS

The structure of water and the band assignments of water in various solvents have been studied extensively in the NIR. The 5155-cm^{-1} (1940-nm) combination band in particular has been very useful for studies and analyses of water in the presence of hydroxyl-containing solvents because it is usually well isolated.

The position and widths of the bands vary predictably with the degree of hydrogen bonding and the basicity of the solvent.[16] Interactions are strongest in dimethyl sulfoxide (DMSO) and weakest in nitromethane, as demonstrated by the position of the 5155-cm^{-1} (1940-nm) peak. The series of solvents includes, in order, DMSO, dioxane, acetone, acetonitrile, and nitromethane, and the peak maximum ranged from 5150 cm^{-1} (1942 nm) to 5270 cm^{-1} (1898 nm).[17] The effect of low levels of ethanol added to water was seen to be qualitatively the same as a temperature effect.[18]

In all solvents, the fraction of free water was greater in both dilute solution and in neat water than it was in intermediate concentrations. Through the analysis of ternary mixtures, the amount of nonbonded water molecules was found to be negligibly small in proton-acceptor solvents.

6.5 WATER IN OTHER MATRICES, INCLUDING GLASSES

The relative position of the water bands has provided a good deal of information to researchers. In studying protein denaturation, for example, the symmetric stretch plus asymmetric stretch combination band provided information on the state of the water during the reaction. The shift from 1410 to 1490 nm indicated an increase in bound water.[19] In studying water sorption on PET film, the same spectral region showed that water was not interacting with the film. The wavelengths of the sub-bands in that region were all lower than those in bulk water.[20] The total water content of starch and cellulose was compared to the "water activity" using the 1450- and 1940-nm bands.[21] It was found that NIR could discriminate between different levels of moisture content at the same activity but not very well between activities at the same moisture content. The 1940-nm band was also used in a cell adhesion study.[22]

Water incorporated into fused silica is of considerable importance because the resulting silanol groups affect the NIR transmission of silica optical fibers and other optical components. Silanols are discussed in the OH section. In mixed glasses, such as those containing borosilicates, aluminosilicates, and so on, associated water OH groups give rise to diffuse absorption bands.[23] The absorption bands of water molecules on silica surfaces have been described by Klier et al.[24]

6.6 IONIC SPECIES IN WATER

There are several ways that ionic species in water affect near-infrared spectra. These affects include (1) a decrease in the concentration of water, (2) charge-dipole interactions between the ions and the water molecules that affect the hydrogen bonding of the water itself, (3) the formation of hydrogen bonds between an oxygen or nitrogen in the ions and water, (4) the presence of OH itself or OH within an ion (such as HCO_3^-), and (5) the presence of another specific NIR-active functional group within the ion (such as acetate's CH and ammonia's NH), and the condition by which some ions produce OH or H through hydrolysis (such as $CO_3^=$).[25] In addition, the color of some ions can extend into the NIR region.

One example of an ionic interaction with water molecules is shown in Figure 6.5. Sodium chloride alone in water can be readily measured in spite of its having no NIR bands of its own because it

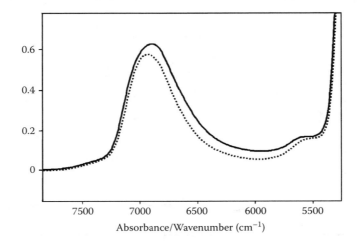

FIGURE 6.5 The effect of dissolved sodium chloride on the water spectrum. The solid curve is pure water, whereas the dotted curve is 20% NaCl.

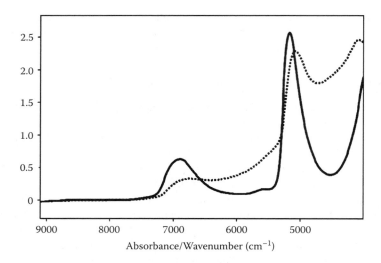

FIGURE 6.6 Spectrum of concentrated HCl (dotted curve) compared to pure water in a 0.5-mm pathlength cell.

affects the spectrum of water by reducing the amount of hydrogen bonding within the bulk water.[26,27] The salt lowers the intensity of the water peak and also shifts it to the left, toward higher frequencies, indicating less hydrogen bonding.

Different ions have somewhat different effects, depending on their size and electronic characteristics. Thermodynamic calculations have indicated that divalent magnesium ions, trivalent aluminum ions, and protons enhance the structural order of the water, whereas monovalent sodium and potassium disrupt it.[28] Differences in the intensities of changes to the near-infrared spectra have been attributed to the charge-to-radius ratios of cations or anions.[29] An example of the effect of a proton is shown in the HCl spectrum in Figure 6.6. Note the shift towards lower wavenumber, indicating an increase in hydrogen bonding. Hydroxide and fluoride anions are said to enhance the structure, whereas chloride, bromide, nitrite, and isocyanate disrupt it.[30] Choppin and Buijs based their assessment of structure effects on the 8330-cm⁻¹ (1200-nm) combination band by calculating the relative intensities of the free, singly bonded, and doubly bonded peaks. Measurements of pH and titration endpoints by NIR based on the effects on the water peaks have been reported.[31]

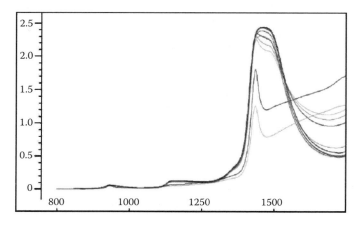

FIGURE 6.7 Sodium hydroxide solutions, 0–50% NaOH. The bottom curve is 50% NaOH, and the top is 100% water.

The ionic hydroxyl group in aqueous solutions has its own specific absorption bands with first and second overtone bands at 7040 cm^{-1} (1421 nm) and 1034 cm^{-1} (967 nm), respectively.[32,33] Also, a broad band centered at about 1100 nm has been attributed to the binding of two water molecules to the hydroxide ion. As the hydroxide concentration is increased above 5 molal, the water peaks at 6900 cm^{-1} (1450 nm) and 1025 cm^{-1} (976 nm) decrease, and the hydroxide ion peaks at 1421 and 967 nm become prominent. This effect is complicated, however, and involves water activity and the ability of the solutes to confine the bulk solvent within hydration spheres. KOH solutions do not behave the same as NaOH solutions, and the effects of the cation on the water absorption need to be accounted for to obtain good measurements of hydroxide. Figure 6.7 illustrates the prominent ionic hydroxyl peak near 7040 cm^{-1} (1421 nm). It is especially noticeable in the 50% solution.

REFERENCES

1. Yamatera, H., Fitzpatrick, B., and Gordon, G., Near-infrared spectra of water and aqueous solutions, *J. Mol. Spectrosc.*, 14, 268–278, 1964.
2. Polyansky, O.L., Zobov, N.F., Viti, S., and Tennyson, J., Water vapor line assignments in the near-infrared, *J. Mol. Spectrosc.*, 189, 291–300, 1998.
3. Buijs, K. and Choppin, G.R., Near-infrared studies of the structure of water. I. Pure water, *J. Chem. Phys.*, 39(8), 2035–2041, 1963.
4. Choppin, G.R. and Violante, M.R., Near-infrared studies of the structure of water. III. Mixed solvent systems, *J. Chem. Phys.*, 56(12), 5890–5898, 1972.
5. Choppin, G.R. and Downey, J.R., Near-infrared studies of the structure of water. IV. Water in relatively nonpolar solvents, *J. Chem. Phys.*, 56(12), 5899–5904, 1972.
6. Ellis, J.W. and Sorge, B.W., The infrared absorption spectrum of water containing deuterium, *J. Chem. Phys.*, 2, 559–564, 1934.
7. Inoue, A., Kojima, K., Taniguchi, Y., and Suzuki, K., Near-infrared spectra of water and aqueous electrolyte solutions at high pressures, *J. Solution Chem.*, 13(11), 811–823, 1984.
8. Buijs, K. and Choppin, G.R., op. cit.
9. Inoue, A., Kojima, K., Taniguchi, Y., and Suzuki, K., Near-infrared spectra of water and aqueous electrolyte solutions at high pressures, *J. Solution Chem.*, 13(11), 811–823, 1984.
10. Buijs, K. and Choppin, G.R., op. cit.
11. Choppin, G.R. and Violante, M.R., op. cit.
12. Inoue, A., Kojima, K., Taniguchi, Y., and Suzuki, K., op. cit.
13. McCabe, W.C., Subramanian, S., and Fisher, H.F., A near-infrared spectroscopic investigation of the effect of temperature on the structure of water, *J. Phys. Chem.*, 74, 4360–4369, 1970.
14. Maeda, H., Ozaki, Y., Tanaka, M., Hayashi, N., and Kojima, T., Near-infrared spectroscopy and chemometrics studies of temperature-dependent spectral variations of water: relationship between spectral changes and hydrogen bonds, *J. Near-Infrared Spectrosc.*, 3, 191–201, 1995.
15. Bayly, J.G., Kartha, V.B., and Stevens, W.H., The absorption spectra of liquid phase H$_2$O, HDO, and D$_2$O from 0.7 μm to 10 μm, *Infrared Phys.*, 3, 211–223, 1963.
16. Choi, Y.-S., Near-infrared spectroscopic investigations of hydrogen-bonding of water in organic solvents and alcohols, and partial molal volume studies of some solutes in H$_2$O and D$_2$O, Ph.D. dissertation, University of South Carolina, 1974.
17. Burneau, A., Near-infrared spectroscopic study of the structures of water in proton acceptor solvents, *J. Mol. Liq.*, 46, 99–127, 1990.
18. Onori, G., Near-infrared spectral study of aqueous solutions of ethyl alcohol. *Il Nuovo Cimento*, 10(4), 387–394, 1988 (in English).
19. Vandermeulen, D.L. and Ressler, N., A near-infrared method for studying hydration changes in aqueous solution: illustration with protease reactions and protein denaturation, *Arch. Biochem. Biophys.*, 205(1), 180–190, 1980.
20. Fukuda, M., Kawai, H., Yagi, N., Kimura, O., and Ohta, T., FTI.R. study on the nature of water sorbed in poly(ethylene terephthalate) film, *Polymer*, 31, 295–302, 1990.
21. Delwiche, S.R., Pitt, R.E., and Norris, K.H., Sensitivity of near-infrared absorption to moisture content versus water activity in starch and cellulose, *Cereal Chem.*, 69(1), 107–109, 1992.

22. Grant, D., Long, W.F., and Williamson, F.B., NIR spectroscopy shows that animal cell adhesion to non-biological solid surfaces may require surface water structuring, in *Making Light Work*, Murray, I. and Cowe, I.A. (Eds.), VCH, Weinheim, Germany, 1992, pp. 636–639.

23. Spierings, G.A.C.M., The near-infrared absorption of water in glasses, *Phys. Chem. Glasses*, 23(4), 101–106, 1982.

24. Klier, K., Shen, J.H., and Zettlemoyer, A.C., Water on silica and silicate surfaces. I. Partially hydrophobic silicas, *J. Phys. Chem.*, 77(11), 1458–1464, 1973.

25. Lin, J., Zhou, J., and Brown, C.W., Identification of electrolytes in aqueous solutions from near-infrared spectra, *Appl. Spectrosc.*, 50(4), 444–448, 1996.

26. Hirschfeld, T., Salinity determination using NIRA, *Appl. Spectrosc.*, 39, 740–743, 1985.

27. Grant, A., Davies, A.M.C., and Bilverstone, T., Simultaneous determination of sodium hydroxide, sodium carbonate, and sodium chloride concentrations in aqueous solutions by near-infrared spectrometry, *Analyst*, 114, 819–822, 1989.

28. Molt, K., Berentsen, S., Frost, V.J., and Niemoeller, A., NIR spectrometry — an alternative for the analysis of aqueous systems, *Spectrosc. Eur.*, 10(3), 16–21, 1998.

29. Lin, J. et al., op. cit.

30. Choppin, G.R. and Buijs, K., Near-infrared studies of the structure of water. II. Ionic solutions, *J. Chem. Phys.*, 39(8), 2042–2050, 1963.

31. Molt, K. et al., op. cit.

32. Heiman, A. and Licht, S., Fundamental baseline variations in aqueous near-infrared analysis, *Anal. Chim. Acta*, 394, 135–147, 1999.

33. Phelan, M.K., Barlow, C.H., Kelly, J.J., Jinguji, T.M., and Callis, J.B., Measurement of caustic and caustic brine solutions by spectroscopic detection of the hydroxide ion in the near-infrared region, *Anal. Chem.*, 61, 1419–1424, 1989.

7 Carbonyls

7.1 ORGANIC CARBONYL COMPOUNDS, OVERTONES

Aldehydes, ketones, esters, anhydrides, acid chlorides, and carboxylic acids show some near-infrared carbonyl-associated bands, as illustrated in Figure 7.1. The C=O stretch is very strong in the mid-infrared, so, even though the first overtone is still in the mid-infrared region, the second overtone may be strong enough to observe in the near-infrared. Although the second overtones are relatively weak and would be overwhelmed if any water were present, there are some anhydrous situations in which it might be useful to analyze carbonyl compounds by near-infrared spectroscopy.

As in the mid-infrared, the position of the C=O varies with the environment of the group. The position or frequency of the fundamental carbonyl C=O stretching vibration is affected by (1) the isotope effects and mass change of substituted groups, (2) bond angles of substituted groups, (3) electronic (resonance and inductive) effects, and (4) interactions of these effects. Substituents with higher mass decrease the C=O stretch frequency; increasing the mass or bond angles of substituents also decreases the fundamental band frequency by up to 40 cm^{-1} at 1715 cm^{-1}. Similarly, decreasing the substituent mass, or bond angles between the carbonyl carbon and its substituents, increases the frequency by 25 cm^{-1} above the nominal 1715-cm^{-1} carbonyl C=O stretch frequency. More electronegative (electron-withdrawing) substituents will increase the carbonyl carbon–oxygen stretch frequency by up to 100 cm^{-1} above the 1715-cm^{-1} nominal frequency. Conjugation of the carbonyl group to aromatic or olefinic groups tends to lower the frequency for both C=O and C=C by 30 to 40 cm^{-1}, depending upon the ring size of the substituent.

In general, the position of the second overtone of the carbonyl of simple noncyclic aliphatic compounds shifts to lower wavelength in the series: acid chlorides, anhydrides, carboxylic acid monomer, lactones, aldehydes, ketones, and esters. Amide carbonyl second overtones are overwhelmed by bands that involve NH stretch and cannot therefore be included in this comparison. An acid chloride second overtone is at 5400 cm^{-1} (1850 nm), propionaldehyde is at 5100 cm^{-1} (1960 nm), acetone is at 5100 cm^{-1} (1960 nm) (with an additional split band at 5260 cm^{-1} [1900 nm]), and ethyl acetate is at 5160 cm^{-1} (1940 nm).

Conjugation with aromatic rings and double bonds, and interactions with halogens and cyclic structures, all affect the band positions, as in the mid-infrared. Also, these second overtones are sometimes split into two as there may be two or more small bands in the vicinity of the calculated second overtone.[1] The second overtone of the C=O of a carboxylic acid appears at about 5260 cm^{-1} (1900 nm) and is particularly clear in a spectrum of perfluorocaproic acid taken in solution, as this acid has no CH absorptions.[2] It can be seen in Figure 7.2, a comparison between octanoic and octadecanoic acids in carbon tetrachloride. With the shorter-chain acid, the acid carbonyl peak is more prominent.

The C–H stretch overtones of aldehydes should be observed at shorter wavenumbers than most compounds as they are in the mid-infrared. They are not very distinctive, however. In the mid-infrared region, the C–H stretch is usually a doublet and involves interaction with an overtone of a bending mode.[3] Therefore, it is not likely to be very strong in the near-infrared. The C–H stretch of benzaldehyde's alkyl is at 5376 cm^{-1} (1860 nm).[4] It is very weak, possibly due to perturbation by a combination band.

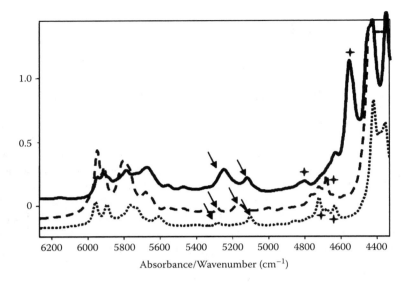

FIGURE 7.1 Comparison of different types of carbonyl compounds. *Note*: The solid curve is neat propionaldehyde, dashed curve is ethyl acetate in CCl$_4$, dotted curve is neat acetone; arrows indicate the carbonyl second
overtone doublets, and stars indicate combination bands involving the carbonyls.

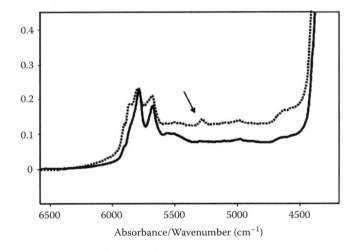

FIGURE 7.2 Illustration of acid carbonyl second overtone; octanoic acid (dotted curve) and octadecanoic
acid (solid).

7.2 ORGANIC CARBONYL COMPOUNDS, COMBINATION BANDS

In addition to the overtones, there are some combination bands that include the C=O stretch.
Combination bands involving C=O and CH in ketones are weak, if they exist at all, because the
two groups do not share a common carbon atom. In aldehydes and formates, however, there is a
carbon atom with both a proton and an oxygen atom attached, which gives rise to a combination band
in the region of 4760 to 4445 cm^{-1} (2100–2200 nm).[5,6] The strong formate ester band at 4650 cm^{-1}
(2150 nm) was assigned to C–H stretch and C=O stretch by a deuteration study.[7]

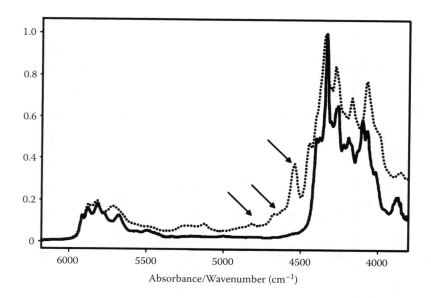

FIGURE 7.3 Pentane (solid curve) and pentanal (dotted curve) compared; combination bands are indicated with arrows and are described in the text.

In one study of short-chain aliphatic aldehydes, the C–H stretch + C=O stretch combination band was given as a peak at 4504 cm⁻¹ (2220 nm) and a 2 × CH bend (in-plane rocking) plus the carbonyl stretch at about 4514 cm⁻¹ (2215 nm).[8] These appear to be one peak in Figure 7.3. Also a peak near 4888 cm⁻¹ (2045 nm) was assigned as 2 × C=O stretch plus 2 × CH bending (out-of-plane wagging),[9] and one at 4748 cm⁻¹ (2106 nm) was assigned as a combination of CH bending (in-plane rocking) and C=O stretching. Figure 7.3 illustrates these absorption peaks. There is also an aldehyde-associated peak near 7850 cm⁻¹, as seen in Figure 7.4, and aldehyde spectra

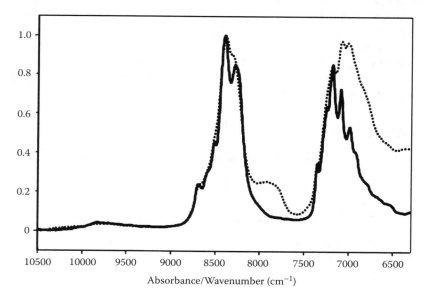

FIGURE 7.4 Pentane (solid curve) and pentanal (dotted curve) in the short wavelength region.

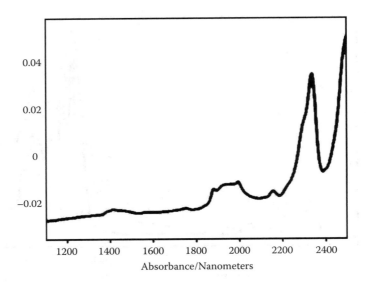

Absorbance/Nanometers

FIGURE 7.5 Reflection spectrum of powdered calcium carbonate.

found in the appendices. This could be a second overtone involving the 2700-cm^{-1} funda-mental CH peak. However, the fundamental C–H stretch absorption is split by Fermi resonance, and therefore, near-infrared peaks are probably also complicated by Fermi resonance.

7.3 CARBONATE

The inorganic ion CO_3 has been analyzed in the near-infrared region in many geological studies. It has two strong bands of particular usefulness near 3920 and 4255 cm^{-1} (2550 and 2350 nm, respectively). The 3920-cm^{-1} band is most likely due to the sum of an infrared-forbidden symmetric stretch and twice the asymmetric stretch ($\nu_1 + 2\nu_3$). The 4255-cm^{-1} band has been assigned as the second overtone of the strong asymmetric stretch (3 ν_3).[10,11] Figure 7.5 shows the strong 4255-cm^{-1} (2350-nm) band clearly.

The band positions shift with environment and are in different positions in calcium carbonate (polymorphs calcite and aragonite), dolomite (calcium magnesium carbonate), sodium carbonate, and so on, alone or in aqueous solutions.

7.4 ISOCYANATES

Table 7.1 provides some band assignments made for the isocyanate group N=C=O.[12]

TABLE 7.1
Carbonyl 1 — Isocyanate Band Assignments

Wavenumber (cm^{-1})	Wavelength (nm)	Assignment
4872	2052	$2 \times \delta$ (NCO) + ν (NCO)
5220	1915	ν (NCO) + ν CH

REFERENCES

1. Ellis, J.W., The near-infrared absorption spectra of some aldehydes, ketones, esters and ethers, *J. Am. Chem. Soc.*, 51, 1384–1394, 1929.
2. Holman, R.T. and Edmondson, P.R., Near-infrared spectra of fatty acids and some related substances, *Anal. Chem.*, 28(10), 1533–1538, 1956.
3. Colthup, N.B., Daly, L.H., and Wiberley, S.E., *Introduction to Infrared and Raman Spectroscopy*, Academic Press, New York, 1990.
4. Srivastava, P.K., Ullas, G., and Rai, S.B., Overtone spectroscopy of benzaldehyde, *Pramana J. Phys.*, 43(3), 231–236, 1994.
5. Kaye, W., Near-infrared spectroscopy, *Spectrochim. Acta*, 6, 257–287, 1954.
6. Holman, R.T. and Edmondson, P.R., op. cit.
7. Powers, R.M., Tetenbaum, M.T., and Tai, H., Determination of aliphatic formates by near-infrared spectrophotometry, *Anal. Chem.*, 34, 1132–1134, 1962.
8. Lucazeau, G. and Sandorfy, C., The near-infrared spectra of some simple aldehydes, *Can. J. Chem.*, 48, 3694–3703, 1970.
9. Kellner, L., Absorption spectra of four aldehydes in the near-infrared, *Proc. R. Soc. London*, A157, 100–113, 1936.
10. Oliver, B.G. and Davis, A.R., Vibrational spectroscopic studies of aqueous alkali metal bicarbonate and carbonate solutions, *Can. J. Chem.*, 51, 698–702, 1972.
11. Gaffey, S.J., Spectral reflectance of carbonate minerals in the visible and near-infrared (0.35–2.55 microns): calcite, aragonite, and dolomite, *Am. Mineralogist*, 71, 151–162, 1986.
12. Brandenbusch, K., Ph.D. thesis, University of Essen, 2001, p. 42.

8 Amines and Amides

8.1 N–H FUNCTIONAL GROUPS

The N–H functional group found in amines and amides has not been studied as extensively as hydroxyls, but it is important because it appears in many natural products, pharmaceuticals, and polymers. N–H participates in hydrogen bonding and therefore behaves differently in various solvents and matrices. The near-infrared region offers a special advantage in the measurement of the primary amine group NH_2 due to a unique combination band.

8.2 AMINES, ALIPHATIC

8.2.1 OVERTONES

As shown in Figure 8.1, primary and secondary amines are distinctly different in the first overtone region near 6600 cm^{-1}. Primary amines have a doublet, and secondary a single peak. Tertiary amines have no peak as they have no N–H functionality and are not shown in the figure. The asymmetric and symmetric NH-stretching peaks occur at 6553 and 6730 cm^{-1} (1625 and 1486 nm), respectively, in butyl amine in carbon tetrachloride. The first overtone of secondary amines has only one band near 6530 cm^{-1} (1530 nm).

Figure 8.1 shows a primary and a secondary amine in dilute carbon tetrachloride solution. The NH-stretch first overtones broaden and shift to longer wavelengths with an increase in concentration, and shift slightly in other solvents. It has been reported that the band position does not change with temperature in the range of 70–120°C in contrast with the hydroxyl group.[1] The symmetric absorption of the –NH_2 group is much more intense (~6 to 7 times) than the asymmetric absorption.[2]

Polar groups next to the amine tend to displace the first overtone. An amine next to phosphorus has its first overtone at 6720 cm^{-1} (1488 nm) and 7652 cm^{-1} (1481 nm).[3]

In the mid-infrared region, the amine NH-stretch absorption is weak relative to primary alcohols, about 1–2 l/mol-cm compared to 50–100 l/mol-cm.[4] However, the intensity of the first overtone of aliphatic amines is of the same order of magnitude as the fundamental. For example, n-butyl amine's first overtone has an absorptivity of 0.6 l/mol-cm compared to 2.4 l/mol-cm for the fundamental. The overtones of amines and hydroxyls are of approximately the same magnitude, and it may be easier to detect an amine in the presence of alcohols in the near-infrared than the mid-infrared.

A single peak that occurs near 9700 cm^{-1} (1035 nm) has been assigned to the second overtone of -NH_2. The third overtone occurs as a doublet at 12,407 and 12,837 cm^{-1} (806 and 779 nm). The absorption of the latter band is also stronger, and its intensity increases with increasing chain length. The fourth overtone appears as a doublet near 15,129 cm^{-1} (661 nm), but one of the pair is of much lower intensity.[5]

8.2.2 COMBINATIONS

The first combination band region of primary amines, which derives from the N–H- stretching modes and the deformation or scissoring of the NH_2 group, has been found to be very useful in quantitative and qualitative analyses. As shown in Figure 8.1, n-butyl amine has a strong doublet in CCl$_4$ near 5000 cm^{-1} that is totally absent in secondary (or tertiary) amines. This feature is not a doublet in all solvents, and its dual nature has been attributed to interaction with the solvents. It is a single peak in benzene and hexane but a double peak in mixtures of benzene and chloroform.[6] The lower frequency peak increases upon addition of more polar solvents, and its position has been used to

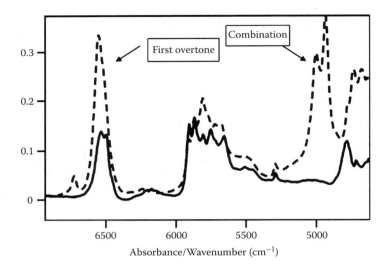

FIGURE 8.1 Primary and secondary aliphatic amines, n-butyl amine (dashed line) and di-isobutyl amine (solid line).

study solvent associations.[7] Sinsheimer and Keuhnelian have interpreted the doublet as being due to a free (higher wavenumber) and a hydrogen-bonded band (lower wavenumber).[8]

A second set of combination bands has been reported at 12,000 and 12,380 cm^{-1} (832 and 808 nm) in secondary amines.[9]

8.3 AMINES, AROMATIC

8.3.1 OVERTONES

The NH-stretching bands of aromatic amines such as aniline show a doublet in the first overtone region, but it is shifted to lower wavelengths relative to aliphatic amines. Figure 8.2 compares the near-infrared spectrum of aniline in CCl$_4$ with that of n-butyl amine. The asymmetric vibration occurs

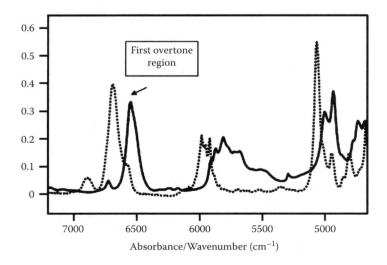

FIGURE 8.2 Aniline (dotted line) compared to n-butylamine (solid line).

TABLE 8.1
Secondary Amine First Overtone Band Intensity Comparisons

Compound	Area of Fundamental $\times 10^9$	Area of First Overtone $\times 10^9$	Ratio of Areas	Position of Overtone (cm^{-1})
Dimethylamine (vapor)	~6	9.5	~0.6	6580
Diethylamine	~3.5	~4.8	~0.7	6471
Morpholine	21	10	2.1	6536
Dibenzylamine	30	6.2	4.9	6490, 6512, 6534
N-benzylamine	183	7.5	24	6751
Diphenylamine	228	5.6	40	6729
Indole	568	11.2	59	6844
Carbazole	550	8.8	62	6826

Note: CCl$_4$ solution. Area in units of cm^2 mol^{-1} sec^{-1}

at 6890 cm^{-1} (1450 nm), and the symmetric band is in the region of 6685 cm^{-1} (1496 nm). In studies of ring-substituted derivatives, Whetsel[10] has investigated the correlation of a substituent electronic nature with band positions and intensities in both NH-stretching and combination modes. The shifts of NH overtone bands may be observed as effects of various solvents and temperature conditions.

As in the case of the aliphatic amines, secondary aromatic amines show only one NH-stretching overtone. For N-butyl aniline in CCl$_4$, for example, the overtone is at 6675 cm^{-1} (1498 nm). The effects of hydrogen bonding on the mechanical anharmonicity of the NH-stretching vibration of N-methylaniline have been reported.[11]

In heterocyclic aromatic amines such as pyrrols, indoles, and carbazoles, there is a first overtone NH-stretching between 6803 cm^{-1} (1470 nm) and 6897 cm^{-1} (1440 nm).[12] Unlike the aliphatic amines, where the intensity of the overtone is approximately the same as the intensity of the fundamental, the aromatic and heterocyclic amines have a much greater absorptivity difference between fundamental and overtone. The overtone intensities do not vary widely, but the intensities of the fundamentals vary by a factor of about 100 from aliphatic to heterocyclic. Table 8.1 lists some of the intensities, given in integrated peak area.

Table 8.2 lists band positions and intensities of some primary aromatic amines in carbon tetrachloride, illustrating the effect of substituents.[13] The table also includes the second overtone

TABLE 8.2
Example Aromatic Amine Band Overtones

Substituent	First Overtone Symmetric (cm^{-1})	First Overtone Symmetric (nm)	First Overtone Asymmetric (cm^{-1})	First Overtone Asymmetric (nm)	Second Overtone (cm^{-1})	Second Overtone (nm)
p-NH$_2$	6656	1502.5	6852	1459.5	9741	1026.5
p-CH$_3$	6683	1496.5	6885	1452.5	9775	1023.0
None	6698	1493.0	6904	1448.5	9794	1021.0
p-Cl	6705	1491.5	6906	1448.0	9813	1019.0
m-Cl	6713	1489.5	6920	1445.0	9818	1018.5
m-NO$_2$	6725	1487.0	6940	1441.0	9852	1015.0
o-OCH$_3$	6700	1492.5	6916	1446.0	9818	1018.5
o-Cl	6705	1491.5	6930	1443.0	9828	1017.5
o-NO$_2$	6791	1472.0	6982	1432.0	9970	1003.0

TABLE 8.3
Example Combination Band Positions of Aromatic N–H Peaks

Substituent	Combination band (cm^{-1})	Combination band (nm)
p-NH$_2$	5049	1980.5
p-CH$_3$	5062	1975.5
None	5072	1971.5
p-Cl	5075	1970.5
m-Cl	5082	1968.0
m-NO$_2$	5095	1963.0
o-OCH$_3$	5057	1977.5
o-Cl	5072	1971.5
o-NO$_2$	5084	1967.0

band positions. Molar absorptivities, generally around 1.3 l/mol-cm for the symmetric band, around 0.15 l/mol-cm for the asymmetric, and 0.03 l/mol-cm for the second overtone, were also provided in the reference.

The self-association of aromatic amines in various solvents and temperatures has been studied extensively and has been summarized by Whetsel.[14] The formation of dimers and higher-order associations has been proposed. Association in carbon tetrachloride occurs and some authors have found chloroform to be a better choice for analyzing amines in general.

8.3.2 COMBINATIONS

The band near 5070 cm^{-1} (1972 nm) is due to a combination of NH stretching and bending. A number of examples of band positions are included in Table 8.3.

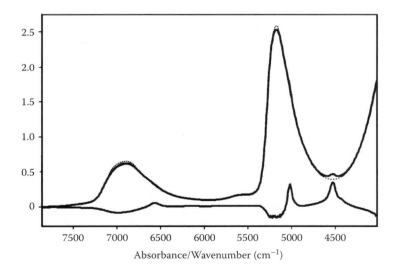

FIGURE 8.3 NIR spectrum of ammonia in water (bottom) compared to water itself (top). The top curve is 5% ammonium hydroxide in water. The dotted curve is pure water, and the bottom curve shows the difference between the two, slightly expanded to accentuate the peaks.

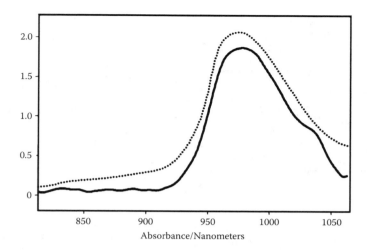

FIGURE 8.4 Low-wavelength region of 10% ammonia in water (solid curve) compared to water (dotted curve).

8.4 AMMONIA

Ammonia, being a weak base, is primarily in the form of NH_3 in water, and its spectrum resembles that of an amine. Ammonia in water shows strong bands at 6520 cm^{-1} (1534 nm), 5000 cm^{-1} (2000 nm), and 4525 cm^{-1} (2210 nm) in addition to water bands.[15] See Figure 8.3. The first overtone near 6520 cm^{-1} appears as a single peak even in gaseous ammonia.[16] In fact, gaseous ammonia has been recommended as a wavelength standard for near-infrared spectrometers. In a 10-cm cell, bands at 6609 cm^{-1} (1513 nm), 5084 cm^{-1} (1967 nm), and 4417 cm^{-1} (2264 nm) were cited.

There is also a sharp second overtone peak at about 9560 cm^{-1} (1046 nm), as shown in Figure 8.4.[17]

8.5 AMIDES, ALIPHATIC AND AROMATIC

8.5.1 Overtones

Near-infrared spectra of primary amides ($RCO-NH_2$), such as formamide, acetamide, and benzamide, have been studied in chloroform solution.[18] Two bands at 6710 cm^{-1} (1490 nm) and 6995 cm^{-1} (1430 nm) have been assigned to first overtones of the asymmetric and symmetric N—H stretching modes.

The assignment of secondary amides has also been studied extensively. A single band between 6711 and 6803 cm^{-1} (1470 to 1490 nm) for N-methylacetamide and other simple amides was attributed to the first overtone NH stretch. Dilute solutions of N-methylacetamide in either carbon tetrachloride or water have a band at 6711 cm^{-1} (1470 nm), whereas more concentrated solutions have a doublet in the 6666–6880-cm^{-1} (1500–1700-nm) range.[19] In CCl$_4$, the bonded doublet appears at concentrations above 1 M, whereas in water the unassociated band is present until about 8 M.

When the fundamental and first overtone NH-stretch absorptions of secondary amides are compared, it is observed that the relative intensities of the free and bound peaks are very different. As seen in Figure 8.5, the overtone of the "free" NH is relatively stronger in the first overtone when compared to the fundamental. This is because hydrogen bonding has increased the anharmonicity of the vibration, resulting in the higher-intensity overtone.

As in amines, the first overtone of OH- and NH-stretching absorptions are comparable in intensity, whereas in the fundamental region the OH is much stronger. A comparison of OH and NH first overtone peak intensities can be seen in Figure 8.6. The figure illustrates the reaction of phenol with an isocyanate to form an amide.

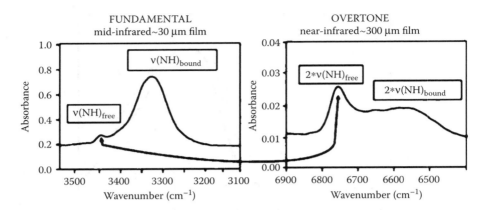

FIGURE 8.5 The effect of hydrogen bonding on the anharmonicity and NH first overtone intensity in an aliphatic polyamide. (From Siesler, H.W., Ozaki, Y., Kawata, S., Heise, H.M., *Near-Infrared Spectroscopy,* Wiley-VCH, Weinheim, 2002. With permission.)

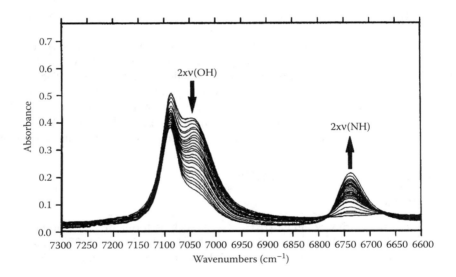

FIGURE 8.6 Hydroxyl first overtone peak decrease and amide NH peak increase during a reaction. (From Dittmar, K. and Siesler, H.W., Fresenius, *J. Anal. Chem.,* 362, 111, 1998. With permission.)

The second overtone of primary amides occurs at 10,110–10,260 cm^{-1} (975 to 989 nm). The second overtone of secondary amides is near 10,194 cm^{-1} (981 nm).

8.5.2 Amides, Combinations

In primary amides, a single combination band of symmetric and asymmetric NH stretching appears at 6805 cm^{-1} (1470 nm). Several bands near 5100 and 4925 cm^{-1} (1960 and 2030 nm) have been assigned to the combinations of NH stretch and amide II and III deformations (these mid-infrared bands are generally considered to be due to different types of coupling of the CNH deformation and

TABLE 8.4
Summary of Urea Absorptions

Mid-Infrared fundamental bands	Wavenumber (cm⁻¹)	Wavelength (nm)
NH asymmetrical stretch	3400	2941
NH symmetrical stretch	3300	3030
C=O stretch (amide I)	1610	6211
NH_2 deformation coupled with C-N stretch (amide II)	1630	6135
C-N stretch coupled with NH_2 deform (amide III)	1467	6817
NH_2 wagging	1150	8696
Near-Infrared bands		
1st overtone NH asymmetrical stretch	6803	1470
1st overtone NH symmetrical stretch	6736	1485
NH asymmetrical + NH symmetrical	6666	1500
Asymmetrical NH stretch + amide II	5025	1990
Symmetrical NH stretch + amide II	4902	2040
Asymmetrical NH stretch + amide III	4808	2080
2 × amide 1 + amide III	4687	2180
Asymmetrical NH stretch + NH_2 rocking	4505	2220

CN stretch). For example, a strong band at 5100 cm⁻¹ (1960 nm) has been assigned to the combination of asymmetric NH stretching with amide II. Two weak bands at 4925 cm⁻¹ (2030 nm) and 4975 cm⁻¹ (2010 nm) have been attributed to the combination of asymmetric NH stretching with amide III and symmetric NH stretching with amide III, respectively.

The near-infrared spectrum of urea, a special case of primary amide, was described by Murray.[20] Urea and thiourea were also discussed by Bala and Ghosh.[21] Table 8.4 provides a summary of its overtone and combination peaks. The fundamental bands listed in the table provide background information to help explain the combinations.

For secondary amides, the combination bands in the range of 4950–5000 cm⁻¹ (2000–2020 nm) and 4695–4740 cm⁻¹ (2110–2130 nm) were assigned to the NH stretch with amide II and NH stretch with amide III, respectively.[22] An additional band at 4630 cm⁻¹ (2160 nm) was attributed to the combination of the first overtone of C=O with amide III ($2v_{C=O}$ + amide III).[23]

The main combination bands of secondary amides are summarized in Table 8.5. These bands are important in the analysis of proteins. As explained by Murray in his chapter on spectral comparisons,[24] two of the combination bands of proteins lie on either side of the OH combination band of carbohydrates. Therefore, the shape of the spectral region near 4760 cm⁻¹ (2100 nm) is

TABLE 8.5
Summary of Secondary Amide Bands in Protein

Band Assignment	Wavenumber	Wavelength
1st overtone NH	6250–6540	1530–1600
NH stretch + amide II	4850	2060
2 × C=O (amide I) + amide III	4590	2180

TABLE 8.6
Secondary Amide Bands in Polyamide 11

Band Assignment	Wavenumber	Wavelength
1st overtone nonbonded NH	6760	1480
1st overtone bonded NH, disordered phase	6600	1515
1st overtone bonded NH, ordered phase	6500	1538
Bonded NH + 2 × amide II	6368	1570
Amide I + 3 × amide II	6256	1598
4 × amide II	6180	1618
Bonded NH + amide I	4970	2012
Bonded NH + amide II	4870	2053
2 × amide II + amide I	4701	2127
Bonded NH + amide III	4586	2183
2 × amide I + amide III	4521	2212

Note: See Table 8.4 for explanation of nomenclature.

indicative of the relative amounts of protein and carbohydrate in food and agricultural products. Secondary amide peaks are also important in the analysis of nylon and many pharmaceuticals.

Wu and Siesler have assigned additional secondary amide bands in their study of the aliphatic polymer polyamide 11 (PA11).[25] Using deuteration, variable temperature, and polarization measurements, the assignments listed in Table 8.6 were made.

The tertiary amides lack the NH absorptions observed in the primary and secondary amides. A band at 4675 cm^{-1} (2139 nm) for N, N-dimethylformamide, at 4650 cm^{-1} (2151 nm) for N, N-diemthylacetamide, and at 4630 cm^{-1} (2160 nm) for N, N-dimethylbenzamide corresponds to the combination band of $2\nu_{C=O}$ + amide III.

8.6 AMIDES, CYCLIC (LACTAMS)

Near-infrared spectra of lactams primarily show bands attributable to nonhydrogen-bonded amides.[26] This is in contrast to linear secondary amides, which do show hydrogen-bonded peaks. The near-infrared region provides a means of distinguishing *cis* (cyclic) and *trans* (linear) amides, especially for hydrogen-bonding studies. Table 8.7 summarizes the band assignments for the unassociated lactam amide bands.

TABLE 8.7
Unassociated Amides Peaks for Gamma-Valerolactam in CCl$_4$

Assignment	Wavenumber (cm^{-1})	Wavelength (nm)
NH stretch + NH bending	4865	2055
NH stretch + amide II	4785	2090
NH stretch + amide III	4715	2120
2 × C=O (amide I) + amide III	4660	2145

REFERENCES

1. Wheeler, O.H., Near-infrared spectra of organic compounds, *Chem. Rev.*, 59, 646, 1959.
2. Conley, R.T., *Infrared Spectroscopy*, Allyn and Bacon, Boston, MA, 1966, chap. 7.
3. Wheeler, O.H., op. cit.
4. Goddu, R.F., Near-infrared spectrophotometry, in *Advances in Analytical Chemistry and Instrumentation*, Vol. 1, Reilly, C.N. (Ed.), Interscience, New York, 1960, pp. 347–424.
5. Wheeler, O.H., op. cit.
6. Lohman, F.H. and Norteman, W.E., Determination of primary and secondary aliphatic amines by near-infrared spectrophotometry, *Anal. Chem.*, 35, 707–711, 1964.
7. Strait, L.A. and Hrenoff, M.K., Near-infrared spectrophotometric study of hydrogen bonding in primary aliphatic amines, *Spectrosc. Lett.*, 8, 165–174, 1975.
8. Sinsheimer, J.E. and Keuhnelian, A.M., Hydrogen-bond studies of the near-infrared combination bands of the butylamines, *Anal. Chem.*, 46, 89–93, 1974.
9. Wheeler, O.H., op. cit.
10. Whetsel, K.B., Near-infrared N-H bands of primary aromatic amines in chloroform solution, *Spectrochim. Acta*, 17, 614–626, 1961.
11. Durocher, G. and Sandorfy, C., A study of the effects of solvent on overtone frequencies, *J. Mol. Spectrosc.*, 22, 347–359, 1967.
12. Russell, R.A. and Thompson, H.W., Vibrational band intensities and the electrical anharmonicity of the NH group, *Proc. R. Soc. London*, A234, 318–326, 1956.
13. Whetsel, K.B., Roberson, W.E., and Krell, M.W., Near-infrared spectra of primary aromatic amines, *Anal. Chem.*, 30(10), 1598–1604, 1958.
14. Whetsel, K.B., op. cit.
15. Murray, I., The NIR spectra of homologous series of organic compounds in International NIR/NIT conference, Hollo, J., Kaffka, K.J., and Goenczy, J.L. (Eds.), Budapest, Hungary, 1987, pp. 13–28.
16. Willard, H.H., Merritt, L.L., and Dean, J.A., *Instrumental Methods of Analysis*, 4th ed., D. Van Nostrand and Company, Princeton, NJ, 1965, p. 153.
17. Baughman, E.H. and Mayes, D., NIR applications in process analysis, *Am. Lab.*, 54–58, October 1989.
18. Krikorian, S.E. and Mahpour, M., The identification and origin of N-H overtone and combination bands in the near-infrared spectra of simple primary and secondary amides, *Spectrochim. Acta*, 29A, 1233–1246, 1973.
19. Klotz, I.M. and Franzen, J.S., The stability of interpeptide hydrogen bonds in aqueous solution, *JACS*, 82, 5241–5242, 1960.
20. Murray, I., op. cit.
21. Bala, S.S. and Ghosh, P.N., Local mode analysis for the overtone spectra of urea and thiourea, *J. Mol. Struct.*, 127, 277–281, 1985.
22. Krikorian, S.E. and Mahpour, M., op. cit.
23. Hecht, K.T. and Wood, D.L., The near-infrared spectrum of the peptide group, *Proc. R. Soc. London*, A235, 174–188, 1956.
24. Murray, I., op. cit.
25. Wu, S. and Siesler, H.W., The assignment of overtone and combination bands in the near-infrared spectrum of polyamide 11, *J. Near-Infrared Spectrosc.*, 7, 65–76, 1999.
26. Krikorian, S.E., The influence of hydrogen-bonded association on the character of the near-infrared spectra of simple cis- and trans-configured secondary amides, *Spectrochim. Acta*, 37A(9), 745–751, 1981.

9 P-H and S-H

9.1 P-H

The first overtone of P-H stretching was found at 5288 cm^{-1} (1891 nm) in a number of organo-phosphorus compounds. It is slightly more intense than the S-H absorption, having a molar absorptivity of about 0.24 l/mol-cm. It is described as being more diffuse and less sharp, however.[1]

The POH group is observed in phosphorothioic acids. The absorption is significantly shifted relative to the hydroxyl in alcohols, as it is in mid-infrared. The near-infrared (NIR) peak appears at about 5241 cm^{-1} (1908 nm).

9.2 S-H

Goddu mentions a weak S-H first overtone band for benzenethiol and 1-butanethiol at 5076–5051 cm^{-1} (1970 to 1980 nm).[2] He cites molar absorptivities of 0.081 and 0.044 l/mol-cm, respectively. This band is probably a nonbonded S-H. It is also weak in the mid-infrared, although strong in the Raman. Although a bonded S-H would be broader and placed at a lower wavenumber (higher wavelength), Williams has indicated that the S-H is not as amenable to hydrogen bonding as the O-H. Liquid thiols (mercaptans) have sharper peaks than liquid alcohols.[3]

The first overtone of the S-H group can be seen in Figure 9.1 as the weak peak near 5050 cm^{-1}.

The phosphorus-thiol group (P-SH) shows a shift from the normal thiol overtone and gives rise to a weak doublet at 5076–5002 cm^{-1} (1970 and 1999 nm).[4]

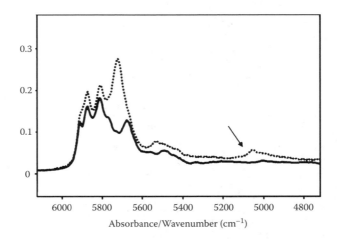

FIGURE 9.1 The S-H first overtone (dotted curve), shown in butane thiol and compared to pentane (solid curve).

REFERENCES

1. McIvor, R.A., Hubley, C.E., Grant, G.A., and Grey, A.A., The infrared spectra of organo-phosphorus compounds, *Can. J. Chem.*, 36, 820–834, 1958.
2. Goddu, R.F., Near-infrared spectrophotometry, in *Advances in Analytical Chemistry and Instrumentation*, Vol. 1, Reilly, C.N. (Ed.), Interscience, New York, 1960, pp. 347–424.
3. Williams, D., The S-H frequency of the mercaptans, *Phys. Rev.*, 54, 504–505, 1938.
4. McIvor, R.A., Hubley, C.E., Grant, G.A., and Grey, A.A., ibid.

10 Carbohydrates

10.1 INTRODUCTION

Carbohydrates include saccharides and polysaccharides, or sugars and starches, and cellulosic or lignin type biomolecules. They consist mostly of aliphatic cyclic groups with attached OH groups and ether linkages. Lignin is representative of aromatic natural product compounds. Thus, the bands normally associated with these functional groups may be observed in the near-infrared (NIR) spectra of carbohydrate molecules.

Table 10.1 is a useful reference indicating the appearance of bands related to starches and sugars. Figure 10.1 shows two of these classic spectra, cornstarch and sucrose. Note the sharp nonbonded OH peak near 1430 nm and the lack of a water peak near 1940 nm in sucrose.

10.2 CELLULOSE AND CELLULOSIC COMPOUNDS

An excellent reference article for background reading is supplied by Blackwell.[1] The article reviews the work in infrared and Raman spectroscopy of cellulosic materials up to 1977, including a detailed table giving calculated and observed band assignments for cellulose I. Table 10.2 shows observed band positions for cellulose I functional groups with corresponding first and second overtone positions between 996 nm and 2500 nm (9091 to 4000 cm^{-1}), and Figure 10.2 shows its spectrum. Bands thought to associate with fiber parameters, such as cellulose and lignin are presented in Table 10.3 and Table 10.4, respectively. Multiple quantitative calibrations have demonstrated that cellulose and lignin are determined using the regions from 4348 to 4237 cm^{-1} (2300 to 2360 nm) and 6042 to 5865 cm^{-1} (1655 to 1715 nm.)[2]

Wingfield[3] discussed the possibilities for NIR detection of cellulose in flour. He summarizes work up until the publication of his book and cites early workers in the field. Marton and Sparks[4] have reported measurements of lignocellulose in the infrared region by using simple linear regression of lignin content vs. the absorbance ratio of 1510/1310 cm^{-1}; these frequencies correspond to a ratio of second overtones occurring near 4529/3929 cm^{-1} (2208/2545 nm). Lignin exhibits the strong presence of many aromatic rings and associated functional groups (see Table 10.4).

Gould et al.[5] reported using an infrared frequency shift of the 2900-cm^{-1} band (3448-nm), which would correspond to a frequency shift at a first overtone band near 5800 cm^{-1} (1724 nm). Mitchell and coworkers[6] determined the acetyl content of cellulose acetate using NIR from 35 to 44.8% acetyl. The spectra were measured in 5-cm cells against a reference solution of pyrrole containing 5% carbon tetrachloride. The maximum absorbance at 6920 cm^{-1} (1445 nm) was plotted against percent acetyl to obtain a calibration line. A standard deviation of 0.22% was indicated by the NIR method. Near-infrared spectroscopy was evaluated for monitoring the acid-catalyzed hydrolysis (thinning) of starch (I) using a univariate calibration model based on the integrated area of the 4400-cm^{-1} (2272-nm) absorption band for carbohydrates.[7]

TABLE 10.1
Bands Associated with Starches and Sugars as C–H- and O–H-Related Bands

Functional Grouping	Nanometers	Wavenumbers
C–H stretching and C–C and C–O–C stretching combination	2500	4000
C–H stretching and CH$_2$ deformation combinations	2280–2330	4283–4386
O–H bending and C–O stretching combination. O–H/C–O Polymeric (.O–H and .C–O)	2100	4762
O–H (2v), .O–H – O–H polymeric (.O–H)	1450	6897
3vO–H, saccharides	1009 and 972	9911 and 10,288 (doublet)

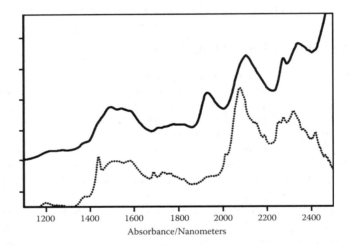

FIGURE 10.1 Corn starch (solid curve) and sucrose (dotted curve).

TABLE 10.2
Cellulose I Associated Bands

Functional Grouping	Nanometers	Wavenumbers
C–H stretching and C–C and C–O–C stretching combination	2500	4000
C–H stretching and C–C stretching combination	2488	4019
3 × (.C–H bending): C–H	2352	4252
C–H (2v CH$_2$ symmetric stretching and δ CH$_2$) combination	2347	4261
C–H stretching and CH$_2$ deformation combination	2280–2330	4283–4386
O–H/C–H Cellulose	2270	4405
O–H polymeric O–H (2v)	2090	4785
O–H stretching and C–O stretching (3v) combination	1820	5495
C–H Methylene (2vC–H)	1780	5618
O–H Polymeric (2vO–H)	1450	6897
O–H (3v)	996	10,040

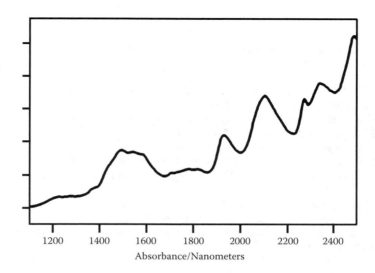

FIGURE 10.2 Cellulose.

TABLE 10.3
Bands Normally Associated with "Fiber" as Cellulosics

Functional Grouping	Nanometers	Wavenumbers
C–H stretching and C–C stretching combination	2488	4019
C–H bending	2352	4252
C–H ($2vCH_2$ symmetric stretching and δCH_2) combination	2347	4261
C–H stretching plus CH_2 deformation combination	2335	4283
O–H stretching plus $2 \times$ C–O stretching combination	1820	5495
CH_2: stretching $(2v)$	1780	5618
O–H polymeric (.O–H): O–H $(2v_S)$, .O–H	1490	6711
O–H $(3v)$ $(-CH_2-OH)$	996	10,040

TABLE 10.4
Bands Normally Associated with "Fiber" as Lignins

Functional Grouping	Nanometers	Wavenumbers
$CC\omega$ + CHv $(11 + 12)$, aromatic ring band assignment	2540	3937 (3935)
C–H stretching and C=O combination	2200	4545
C–H $(2v)$, ArC–H: C–H aromatic associated C–H	1685	5935
C–H $(2vCH_2$ and $\delta CH_2)$ combination: C–H methylene C–H, associated with linear aliphatic $R(CH_2)_N R$	1410	7092
C–H combination, aromatic associated C–H	1417	7057
C–H $(3v)$, .HC=CH	1170	8547
C–H $(4v)$, aromatic associated C–H	876	11,655

REFERENCES

1. Blackwell, J., Infrared and Raman spectroscopy of cellulose, in Arthur, J.C., Jr. (Ed.), *Cellulose Chemistry and Technology*, ACS Symposium Series No. 48, American Chemical Society, 1977, p. 206, chap. 14.
2. Workman, J., Applications of NIR to natural products, *Handbook of Organic Compounds*, Academic Press, Boston, MA, 2001, pp. 170–182, chap. 15.
3. Wingfield, J., NIR detection of cellulose as a milling control parameter, *Association of Operative Millers-Bulletin*, November 1979, pp. 3769–3770.
4. Marton, J. and Sparks, H.E., *TAPPI*, 50, 363–368, 1967.
5. Gould, J.M., Greene, R.V., and Gordon, S.H., Book of abstracts, *188th National Meeting of the American Chemical Society*, Philadelphia, PA, August 1984, American Chemical Society, Washington, D.C., 1984.
6. Mitchell, J.A., Bockman, C.D., Jr., and Lee, A.V., Determination of acetyl content of cellulose acetate by near-infrared spectroscopy, *Anal. Chem.*, 29, 499–502, 1957.
7. Chung, H. and Arnold, M., Near-infrared spectroscopy for monitoring starch hydrolysis, *Appl. Spectrosc.*, 54(2), 277–283, 2000.

11 Amino Acids, Peptides, and Proteins

11.1 PROTEINS

Protein is measured in the near-infrared (NIR) region as its associated functional groups, such as amides and various C–H functional groups. For example, key band locations associated with proteins are found at 10,277 to 9804 cm^{-1} (973 to 1020 nm) as N–H stretch second overtone; 6667 to 6536 cm^{-1} (1500 to 1530 nm) as N–H stretching first overtone; and 4878 to 4854 cm^{-1} (2050 to 2060 nm) representing N–H-stretching combinations. The 4613- to 4587-cm^{-1} (2168 to 2180-nm) region is associated with the N–H bend second overtone and C=O stretch/N–H in-plane bending/C–N stretch combination bands. See Chapter 8 for a more complete description of amide bands.

Hermans and Scheraga[1] made measurements of the backbone peptide hydrogen bond system (NH group) using NIR measurements in the extended spectral region from 14,286 to 2,857 cm^{-1} (700–3500 nm). The authors describe the use of NIR to distinguish between hydrogen-bonded and non-hydrogen-bonded N–H and O–H groups. Molecules examined include methanol, aniline, and poly-γ-benzyl-L-glutamate.

Elliott and Ambrose[2] identified absorption bands for polypeptides and proteins at approximately 3505 cm^{-1} (2853 nm) as N–H-associated stretching bands and at 4825 cm^{-1} (2073 nm) in the overtone region. The main result of their work was to demonstrate that a band at 4840 cm^{-1} (2066 nm) is useful for distinguishing the presence of extended vs. folded configurations for polypeptides and proteins even in the presence of liquid-phase water. The band near 4824 cm^{-1} (2073 nm) was classified earlier by Glatt and Ellis[3] on work in nylon as a combination band of the N–H deformation and stretching modes.

Hecht and Wood[4] described in a detailed study the band assignments for peptides. The authors identified the bands for a porcupine quill spectrum as shown in Table 11.1.

A summary of all bands thought to be associated with protein as amides is presented in Table 11.2. Multiple quantitative calibrations for protein in the agricultural literature demonstrate the area from 4655 to 4545 cm^{-1} (2148 to 2200 nm) as used for protein determination.[5]

Fraser and MacRae[6] have reported important absorption bands for natural product proteins and nylon polyamide at 4870 cm^{-1} (2188 nm) resulting from a combination of the peptide absorptions at 3305 cm^{-1} (3026 nm) and 1540 cm^{-1} (6494 nm). For feather shafts, the authors report absorption bands at 4970 cm^{-1} (2012 nm) and 5040 cm^{-1} (1984 nm). A shoulder at 5040 cm^{-1} (1984 nm) was reported as resulting from side chain amide groups. For beta-keratin, the paper reports important absorption bands at 4600 cm^{-1} (2174 nm), 4850 cm^{-1} (2062 nm), and 4970 cm^{-1} (2012 nm). The authors were interested in studying the amorphous components of naturally occurring protein structures.

11.2 PROTEIN STRUCTURE

The most pronounced changes for the thermal unfolding of RNase A are observed from 4820 to 4940 cm^{-1}. The strong N–H combination band found at 4867 cm^{-1} in the spectrum of native RNase A shifts to 4878 cm^{-1} upon thermal unfolding. The thermal unfolding of RNase A begins with some changes in β-sheet structure, followed by the loss of α-helical structures, and then ending with the unfolding of the remaining β-sheets.[7] Fourier transform near-infrared (FT-NIR) spectra have been measured for

TABLE 11.1
Band Assignments for Peptide Groups as Delineated by K.T. Hecht and D.L. Wood

Reported Band Assignments	Wavenumber	Wavelength (nm)
C–H stretch + CH deformation	3950–4400	2532–2270
2 × stretch C=O + peptide group mode	4590	2180
N–H stretch + peptide group mode	4850	2060
OH stretch + OH deformation of water	5135	1947
2 × C–H stretch	5700–5850	1750–1710
2 × NH stretch	6500	1540
2 × OH stretch + deformation of water	6900	1450
3 × C–H stretch	8100	1235

TABLE 11.2
Absorption Bands Associated with "Proteins" as Amides

Functional Grouping	Nanometers	Wavenumbers
N–H/C–N/N–H [bonded NH stretching and amide III (C–N stretching/N–H in-plane bending) combination] from polyamide 11	2183	4586
N–H (3δ)	2180	4587
N–H/C–N/C=O [2 × amide I ($2v_s$C=O stretching) and amide III deformation (C–N stretching/N–H in-plane bending) combination] for urea	2180	4687
N–H/C–N/C=O [2 × amide I ($2v_s$C=O stretching) and amide III deformation (C–N stretching/N–H in-plane bending) combination] for secondary amides in proteins	2180	4590
N–H/C–N [v_sN–H asymmetric and amide III deformation (C–N stretching/N–H in-plane bending) combination] for urea	2080	4808
N–H [vN–H and amide II deformation (N–H in-plane bending) combination] for secondary amides in native RNase A	2075	4820
N–H (3δ) and N–H stretching combination	2060	4854
N–H [vN–H and amide II deformation (N–H in-plane bending) combination] for secondary amides in proteins	2060	4850
$CONH_2$ combination of amide A and amide II	2060	4855
N–H stretching and C=O stretching (amide I) combination	2055	4866
N–H (vN–H and δN–H combination) for gamma-valerolactam	2055	4865
N–H combination band found in the spectrum of native RNase A (C=O amide I band)	2055	4867
$CONH_2$ specifically due to peptide β-sheet structures	2055	4865
N–H stretching and C=O stretching (amide I) combination band in the spectrum of native RNase A	2055	4867
N–H from $CONH_2$ as thermal unfolding of RNase A protein in aqueous solution (assigned to an N–H combination band)	2055–2050	4867–4878
N–H in-plane bend and C–N stretching and N–H in-plane bend combination	2050	4878
N–H native RNase A combination band at 4867 cm^{-1} shifting due to thermal unfolding (C=O amide I band)	2050	4878
N–H stretching and C=O stretching (amide I) combination band observed in the thermal unfolding observed in native RNase A	2050	4878
N–H [vN–H symmetric and amide II deformation (N–H in-plane bending) combination] for primary amides	2040	4902
C=O (3v), $C=ONH_2$	2030	4926

TABLE 11.2 (CONTINUED)
Absorption Bands Associated with "Proteins" as Amides

Functional Grouping	Nanometers	Wavenumbers
N–H [vN–H asymmetric and amide III deformation (C–N stretching/N–H in-plane bending) combination] for primary amides	2030	4925
N–H/C=O [bonded NH stretching and amide I ($2v_sC=O$ stretching) combination] of native RNase A	2024	4940
N–H [vN–H symmetric and amide III deformation (C–N stretching/N–H in-plane bending) combination] for primary amides	2010	4975
N–H [vN–H asymmetric and amide III deformation (C–N stretching/N–H in-plane bending) combination] for primary amides	1990	5025
N–H stretching (asymmetric) and N–H in-plane bending combination	1980	5051
$CONH_2$ specifically due to peptide β-sheet structures	1691–1688	5915–5925
N–H (2v), .CONHR	1490	6711
N–H (2v), .$CONH_2$	1483	6743
N–H (2v), .CONHR	1471	6798
N–H (vN–H asymmetric and vN–H symmetric combination) for primary amides	1470	6805

various polypeptides and proteins with different secondary structures in order to identify an NIR marker band for the protein and polypeptide structures. Comparison between FT-NIR and FT–mid-infrared spectra has shown a correlation between the frequency of a band near 4855 cm^{-1}, assignable to a combination of amide A and amide II, and that of a mid-infrared band near 3300 cm^{-1}, assigned to amide A (N–H stretch).[8] The NIR spectra of various proteins (bovine serum albumin, lysozyme, ovalbumin, γ-globulin, β-lactoglobulin, myoglobin, cytochrome c) have been investigated for potential measurement of protein secondary structure. The spectra of proteins in aqueous solutions and as freeze-dried solids indicated α-helix information present at 4090, 4365–4370, 4615, and 5755 cm^{-1}; and β-sheet information at 4060, 4405, 4525–4540, 4865, and 5915–5925 cm^{-1}.[9]

Ovalbumin (OVA) has been studied in acidified, aqueous solutions using two-dimensional (2-D) FT-NIR correlation spectroscopy. This technique demonstrates a significant change in bands when the molecule pH moves from 5.4 to 3.6. A band near 4265 cm^{-1} assigned to a symmetric methylene (CH_2)-stretching mode and a CH_2-bending mode of side chains observed at pH 5.4 disappears completely in the synchronous spectrum at pH 3.6. A band near 4600 cm^{-1} assigned to a combination of amide B (a Fermi resonance band relating to the first overtone of the carbonyl and the N–H stretch) and amide II shifts downward with significant broadening between pH 3.0 and 2.4. A broad band at around 6950 cm^{-1} that was assigned to free water and bound water with weak hydrogen bonds becomes very weak in the synchronous spectrum at pH 2.6, while broad auto peaks around 6450 cm^{-1} suddenly appear that are due to bound water with several hydrogen bonds and the first overtone of an NH-stretching mode of the amide groups of OVA.[10] FT-NIR spectra have been measured for several globular proteins, and a band near 4525 cm^{-1} seems to be highly associated with β-sheet structure. Low-wavenumber NIR bands from the 4500–4000-cm^{-1} region were considered to reflect the amino acid composition of proteins.[11]

REFERENCES

1. Hermans, J. and Scheraga, H.A., Structural Studies of Ribonuclease. IV. The Near-Infrared Absorption of the Hydrogen-Bonded Peptide NH Group, presented before the Division of Biological Chemistry, *138th Meeting of the American Chemical Society*, New York.
2. Elliott, A. and Ambrose, J., Evidence of chain folding in polypeptides and proteins, in *Discuss. Faraday Soc.*, Vol. 9, 246–251, 1950.

3. Glatt, L. and Ellis, J.W., *J. Chem. Phys.*, 16, 551, 1948.
4. Hecht, K.T. and Wood, D.L., The near-infrared spectrum of the peptide group, *Proc. R. Soc. London, Series A, Mathematical and Physical Sciences*, 235(1201), 174–188, April 24, 1956.
5. Workman, J., Applications of NIR to natural products, *Handbook of Organic Compounds*, Academic Press, Boston, MA, 2001, pp. 174–176, chap. 15.
6. Fraser, R.B.D. and MacRae, T.P., Hydrogen-deuterium reaction in fibrous proteins. I, *J. Chem. Phys.*, 29(5), 1024–1028, 1958.
7. Schultz, C.P., Fabian, H., and Mantsch, H.H., Two-dimensional mid-infrared and near-infrared correlation spectra of ribonuclease A: using overtones and combination modes to monitor changes in secondary structure, *Biospectroscopy*, 4(5, Suppl.), S19–S29, 1998.
8. Liu, Y., Cho, R.-K., Sakurai, K., Miura, T., and Ozaki, Y., Studies on spectra/structure correlations in near-infrared spectra of proteins and polypeptides. Part I: A marker band for hydrogen bonds, *Appl. Spectrosc.*, 48(10), 1249–1253, 1994.
9. Izutsu, K.-I., Fujimaki, Y., Kuwabara, A., Hiyama, Y., Yomota, C., and Aoyagi, N., Near-infrared analysis of protein secondary structure in aqueous solutions and freeze-dried solids, *J. Pharm. Sci.*, 95(4), 781–789, 2006.
10. Murayama, K. and Ozaki, Y., Two-dimensional near-infrared correlation spectroscopy study of molten globule-like state of ovalbumin in acidic pH region: simultaneous changes in hydration and secondary structure, *Biopolymers*, 67(6), 394–405, 2002.
11. Miyazawa, M. and Sonoyama, M., Second derivative near-infrared studies on the structural characterization of proteins, *J. Near-Infrared Spectrosc.*, 6(1–4), A253–A257, 1998.

12 Synthetic Polymers and Rubbers*

12.1 INTRODUCTION

Near-infrared (NIR) is routinely used to qualify monomers prior to polymerization reactions. It is used to measure the kinetics of polymer onset and can be used to detect end-point completion and initiator compound levels in polymerization reactions. NIR spectro- scopy can also be used to sort polymers and to control the quality of incoming raw monomers and finished polymeric materials. Molecular spectroscopy using NIR and IR measurement techniques is often used for competitive analysis and to determine thermal or photo-induced oxidation or degradation reactions in polymers. In general, NIR spectroscopy is valuable for polymer identification, characterization, and quantitation. NIR spectroscopy can be completed for *in situ* process applications where no sample preparation is a requirement, and where rugged optical systems are a necessity. Some of the earliest work in applying IR and NIR spectroscopy to polymer characterization is found in References 1 to 11.

12.2 INTERPRETIVE SPECTROSCOPY OF ORGANIC COMPOUND SPECTRA FOR POLYMERS AND RUBBERS IN THE NIR REGION

Polymers and rubbers exhibit characteristic band positions that will vary depending upon the molecular structure and the associated or attached chemical groups. Table 12.1 briefly demonstrates the positions for the major chemical groups encountered in analysis of polymer and rubber materials. The accompanying text describes the individual molecular vibrations and band assignments in much greater detail.

12.3 POLYMERS

Polymers are usually large molecules consisting of repeating units or monomers. Natural polymers exist, such as starches or polysaccharides. Synthetic polymers are commonly termed plastics and are used for many commonly used materials and products. Polymers vary in molecular formula and molecular weight due to variation in the number or repeating units. Polymer backbone structures often have attached molecular groups. The molecular arrangement of these groups determines the stereo-chemical configuration for any given polymer. If all the attached groups are in the identical position along the polymer backbone chain, the polymer is in an *isotactic* configuration. If the attached groups alternate in their attached positions with a regular pattern, the *syndiotactic* configuration is ascribed. When attached groups are randomly attached to the polymer backbone, the polymer is said to have an *atactic* configuration. The isotactic configuration represents the most crystalline (rigid) of the configuration types.

Copolymers involve the use of two or more monomer types into a single backbone structure to achieve specific material performance properties. Structures are classified as *alternating*

---

* Figures in Chapter 12 are reprinted with permission: Workman J., Near infrared spectroscopy of polymers and rubbers. Meyers, R.A. (Ed.). In *Encyclopedia of Analytical Chemistry,* John Wiley & Sons Ltd., Chichester, 2000, pp. 7828–7856.

TABLE 12.1
C–H, N–H, and O–H Stretch Absorption Bands for Specific Long-Wavelength NIR (1100–2500 nm) Functional Groups (1st (2v) through 4th (2v) C–H-Stretching Overtones)

Structure	Bond Vibration	Location of (2v) 1st Overtone	Location of (3v) 2nd Overtone	Location of (4v) 3rd Overtone	Location of (5v) 4th Overtone
R-OH (Alcohols)	O–H stretching	6981 cm^{-1} 1410–1455 nm	10,471 cm^{-1} 940–970 nm	—	—
ArOH (Phenols)	O–H stretching	6918 cm^{-1} 1421–1470 nm	10,417 cm^{-1} 940–980 nm	—	—
Starch	O–H stretching	6892 cm^{-1} 1451 nm	10,341 cm^{-1} 967 nm	—	—
Urea	Symmetric N–H stretching	6849 cm^{-1} 1460 nm	10,173 cm^{-1} 973–993 nm	—	—
HOH (Water)	O–H stretching	6838 cm^{-1} 1440–1485 nm	10,417 cm^{-1} 960 nm	—	—
CONHR (Secondary amides)	N–H stretching	6793 cm^{-1} 1472 nm	10,194 cm^{-1} 981 nm	—	—
CONH2 (Primary amides)	N–H stretching	6787 cm^{-1} 1463–1484 nm	10,183 cm^{-1} 975–989 nm	—	—
Cellulose	O–H stretching	6711 cm^{-1} 1490 nm	10,070 cm^{-1} 993 nm	—	—
ArNH$_2$ (Aromatic amines)	N–H stretching	6698 cm^{-1} 1493 nm	10,050 cm^{-1} 995 nm	—	—
NH (Amines, general)	N–H stretching	6667 cm^{-1} 1500 nm	10,000 cm^{-1} 1000 nm	—	—
Protein	N–H stretching	6618 cm^{-1} 1511 nm	9930 cm^{-1} 1007 nm	—	—
ArCH (Aromatics)	C–H stretching	5945 cm^{-1} 1680–1684 nm	8749 cm^{-1} 1143 nm	11,429 cm^{-1} 875 nm	13,986 cm^{-1} 715 nm
CH$_2$–CH$_2$. (Methylene)	C–H stretching	5792 cm^{-1} 1726–1727 nm	8251 cm^{-1} 1212 nm	10,770 cm^{-1} 927–930 nm	13,115 cm^{-1} 761–764 nm
CH$_3$ (Methyl)	C–H stretching	5706 cm^{-1} 1743–1762 nm	8386 cm^{-1} 1191–1194 nm	10,953 cm^{-1} 911–915 nm	13,396 cm^{-1} 745–748 nm
CH$_3$ (Methyl)	C–H combination	4914 cm^{-1} 2020–2050 nm	7369 cm^{-1} 1347–1367 nm	—	—
>C=O (Carbonyl)	>C=O stretching	3431 cm^{-1} 2915 nm	5175 cm^{-1} 1920–1945 nm	6944 cm^{-1} 1420–1460 nm	8658 cm^{-1} 1140–1170 nm

Source: From Workman, J. and Springsteen, A. (Eds.), *Applied Spectroscopy: A Compact Reference for Practitioners,* Academic Press, Boston, 1998, pp. 46–47.

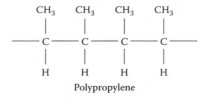

FIGURE 12.1 Polypropylene is a polyolefin comprised of repeating olefin monomer units.*

(-A-B-A-B-A-B-),* *random* (-A-A-B-A-B-B-A-), as *block* (-A-A-B-B-B-B-A-A-), and as *graft* (-A-A-A-A-A-A-A-A-).

B   B   B
B   B   B

Polymerization involves the reaction of the monomer building blocks into polymers. The polymerization reaction types involve *addition* reactions and *condensation* reactions. Addition reactions typically involve the use of ethene to form polyethene and polyethylene. Condensation reactions typically involve different reaction products reacting to form a heteropolymer and a small molecular by-product. The reaction of 1,6-diaminoethane and hexanedioic acid to form nylon and water is a classic example. Polymers made from one monomer type are termed *homopolymers*, and those formed with two different monomers are referred to as *copolymers*.

A few example polymers are shown in the following figures (Figure 12.1, Figure 12.2, Figure 12.3, Figure 12.4, Figure 12.5, Figure 12.6).

Typical polymer spectra exhibit similar NIR bands and positions, which are represented in Table 12.2.

12.4 RUBBERS

Rubber or elastomer is a natural or synthetic material having the ability to undergo deformation under the influence of an applied force and regain its original shape once the applied force is removed. Natural rubbers are polymeric materials originally obtained from the rubber tree (*Hevea brasiliensis*). The sap of this tree is a latex material that is processed by coagulation and drying the sap, and then vulcanizing and adding filler compounds. The basic natural rubber compound is isoprene, which contains the repeating unit of $-CH_2C(CH_3):CHCH_2-$. Synthetic rubbers exist such as nitriles, butadienes, neoprene, butyl rubbers, polysulfide rubbers, PVC, and silicone rubbers, whose structures and NIR spectra are shown in Figure 12.7, Figure 12.8, Figure 12.9, and Figure 12.10.

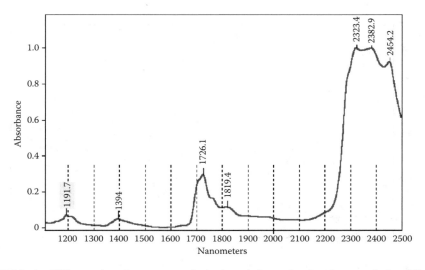

FIGURE 12.2A An NIR wavelength spectrum of atactic poly(propylene) at low resolution (32 cm^{-1}).

---

* Figures 12.1, 2A, 3, 4, 5, 6A, 7, 8A, and 10A are from Meyers, A. (Ed.), *Encyclopedia of Analytical Chemistry*, John Wiley & Sons, Chichester, 2002, pp. 7828–7856. Used with permission.

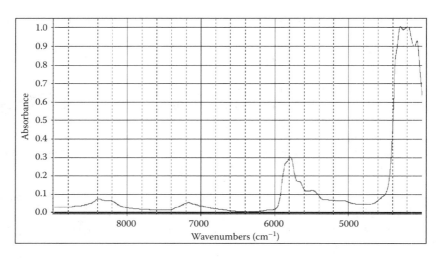

FIGURE 12.2B An NIR wavenumber spectrum of atactic poly(propylene) at low resolution (32 cm⁻¹).

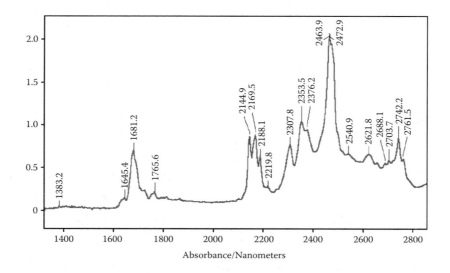

FIGURE 12.3 Structure of styrene monomer units found in polystyrene.

FIGURE 12.4 NIR spectrum of crystalline poly(styrene) at 16-cm⁻¹ resolution.

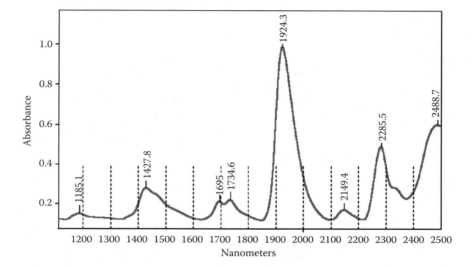

FIGURE 12.5 Two examples of water-soluble polymers as polyacrylates.

FIGURE 12.6A NIR wavelength spectrum of poly(acrylic acid) at 32-cm^{-1} resolution.

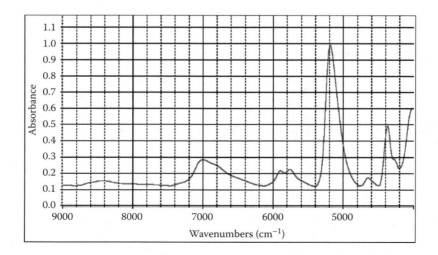

FIGURE 12.6B NIR wavenumber spectrum of poly(acrylic acid) at 32-cm^{-1} resolution.

TABLE 12.2
Spectral Correlation Chart for Example Polymers

Polymer	Band Locations	Band Assignment
Poly(propylene)	9930 cm^{-1} (1192 nm)	Asymmetric methyl (C−H) stretch (3v) 2nd overtone
	8197 cm^{-1} (1220 nm)	Asymmetric methylene (C−H) stretch (3v) 2nd overtone
	7174 cm^{-1} (1394 nm)	Methyl and methylene (C−H) combination
	5882 cm^{-1} (1700 nm)	Asymmetric methyl (C−H) stretch (2v) 1st overtone
	5794 cm^{-1} (1726 nm)	Asymmetric methylene (C−H) stretch (2v) 1st overtone
	5495 cm^{-1} (1820 nm)	Symmetric methyl (C−H) stretch (2v) 1st overtone
	4305 cm^{-1} (2323 nm)	C−H bend (3δ) 2nd overtone
	4196 cm^{-1} (2383 nm)	C−H stretch and C−C stretching combination
	4075 cm^{-1} (2454 nm)	C−H combination band
Poly(styrene)	8757 cm^{-1} (1142 nm)	Aromatic (C−H) stretch (3v) 2nd overtone
	5938 cm^{-1} (1684 nm)	Aromatic (C−H) stretch (2v) 1st overtone
Poly(acrylic acid)	8439 cm^{-1} (1185 nm)	Asymmetric methylene (C−H) stretch (3v) 2nd overtone
	7003 cm^{-1} (1428 nm)	O−H stretch (2v) 1st overtone
	5900 cm^{-1} (1695 nm)	Asymmetric methyl (C−H) stretch (2v) 1st overtone
	5764 cm^{-1} (1735 nm)	Asymmetric methylene (C−H) stretch (2v) 1st overtone
	5198 cm^{-1} (1924 nm)	O−H stretch (2v) 1st overtone + C=O stretch (2v) 2nd overtone + O−H stretch/HOH deformation combination + O−H bend (3δ) 2nd overtone
	4653 cm^{-1} (2149 nm)	C−H stretch/C=O stretch combination + symmetric C−H deformation
	4374 cm^{-1} (2286 nm)	C−H stretch + CH2 deformation + C−H bend (3δ) 2nd overtone
	4018 cm^{-1} (2489 nm)	C−H stretch + C−C stretch + C−O−C stretch comb

FIGURE 12.7 Typical repeating units found in natural rubbers (*trans*- (left) and *cis*- (right) forms).

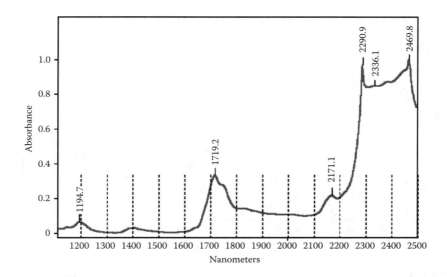

FIGURE 12.8A NIR wavelength spectrum of styrene-isoprene-styrene at 32-cm^{-1} resolution.

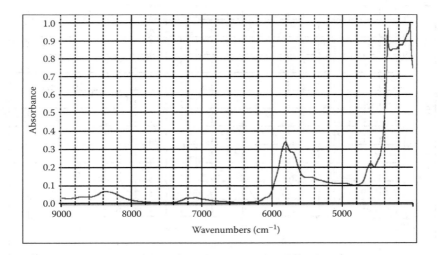

FIGURE 12.8B NIR wavenumber spectrum of styrene-isoprene-styrene at (32-cm^{-1}) resolution.

$$\left[\begin{array}{c} CH_3 \\ | \\ -Si-O- \\ | \\ CH_3 \end{array} \right]_n$$

FIGURE 12.9 Silicone rubbers (polysiloxane). *Note:* The alkyl group is often substituted with a variety of different groups.

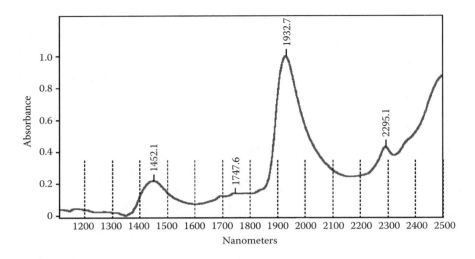

FIGURE 12.10A NIR wavelength spectrum of silicone (DMS) at low (32-cm^{-1}) resolution.

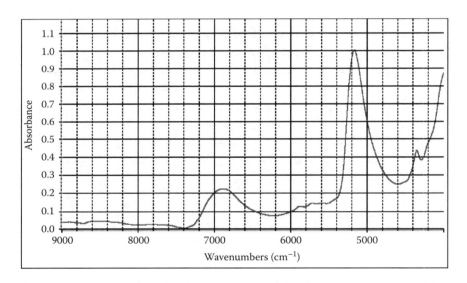

FIGURE 12.10B NIR wavenumber spectrum of silicone (DMS) at low (32-cm^{-1}) resolution.

TABLE 12.3

Spectral Correlation Charts for Model Rubber Compounds

Rubber	Band Location	Description
Styrene-isoprene-styrene	8368 cm^{-1} (1195 nm)	Methyl (C−H) stretch (3v) 2nd overtone
	7143 cm^{-1} (1400 nm)	C−H stretch mode combination
	5817 cm^{-1} (1719 nm)	Asymmetric methylene (C−H) stretch (2v) 1st overtone + aromatic C−H stretch (2v) 1st overtone
	4606 cm^{-1} (2171 nm)	C−H bend (3v) 2nd overtone
	4365 cm^{-1} (2291 nm)	C−H stretch + CH$_2$ deformation combination
	4281 cm^{-1} (2336 nm)	C−H stretch + CH$_2$ deformation combination
	4049 cm^{-1} (2470 nm)	C−H stretch + C−C stretch + C−O−C stretch combination
Silicone (dimethyl siloxane)	6887 cm^{-1} (1452 nm)	Si−OH stretch (2v) 1st overtone
	5721 cm^{-1} (1748 nm)	Methyl (C−H) stretch (2v) 1st overtone
	5173 cm^{-1} (1933 nm)	Si−O−H stretch + Si−O−Si deformation combination
	4357 cm^{-1} (2295 nm)	C−H bend (3v) 2nd overtone

Typical rubber spectra exhibit similar NIR bands and positions, which are represented in Table 12.3.

REFERENCES

1. Staudinger, H., *Ber. Bunsenges. Phys. Chem.*, 53, 1073, 1920.
2. Mark, H. and Raff, R., *Z. Phys. Chem. B*, 31, 275, 1936; Mark, H. and Saito, G., *Mh. Chem.*, 68, 237, 1936.
3. Kaye, W., Near-infrared spectroscopy: a review. I. Spectral identification and analytical applications, *Spectrochim. Acta*, 6, 257–287, 1954.
4. Goddu, R.F., Near-infrared spectrophotometry, *Adv. Anal. Chem. Instrum.*, 1, 347–424, 1960.
5. Whetsel, K.B., Near-infrared spectrophotometry, *Appl. Spectrosc. Rev.*, 2(1), 1–67, 1968.
6. Stark, E., Luchter, K., and Margoshes, M., Near-infrared analysis (NIRA): a technology for quantitative and qualitative analysis, *Appl. Spectrosc. Rev.*, 22(4), 335–399, 1968.
7. Schrieve, G.D., Melish, G.G., and Ullman, A.H., The Herschel-infrared — a useful part of the spectrum, *Appl. Spectrosc.*, 45, 711–714, 1991.
8. Kradjel, C. and McDermott, L., NIR analysis of polymers, in *Handbook of Near-Infrared Analysis*, Burns, D. and Ciurczak, E. (Eds.), Marcel Dekker, New York, 1992.
9. Hummel, D.O., *Infrared Spectroscopy of Polymers*, Wiley-Interscience, New York, 1966.
10. Zbinden, R., *Infrared Spectroscopy of High Polymers*, Academic Press, New York, 1964.
11. Henniker, J.C., *Infrared Spectroscopy of Industrial Polymers*, Academic Press, New York, 1967.
12. Workman, J. and Springsteen, A. (Eds.), *Applied Spectroscopy: A Compact Reference for Practitioners*, Academic Press, Boston, 1998, pp. 46–47.

13 History of Near-Infrared (NIR) Applications

13.1 HISTORY OF NIR FOR INDUSTRIAL CHEMICALS

The application of NIR spectroscopy to the study of basic chemicals began in the 1800s. Following Herschel's discovery of the phenomenon of radiation beyond the visible region,[1] Abney and Festing recorded spectra of organic compounds photographically in 1881 and noted that they were related to the presence of hydrogen atoms.[2] In 1896, Donath[3] measured some organic compounds. In 1899–1900, Luigi Puccianti at the University of Pisa used a single quartz prism to study 15 hydrocarbon compounds.[4] In 1904–1905, William W. Coblentz, working at the Carnegie Institute in Washington, D.C.,[5,6] measured the spectra of a number of compounds including benzene and chloroform from 800–2800 nm using a quartz prism and radiometer measured with the aid of a telescope.[7] In 1922, Joseph W. Ellis at the University of California developed a recording spectrograph, measured a large number of organic compounds, and made many fundamental band assignments.[8] Other early work on specific chemicals included that done at the U.S. Department of Agriculture by Liddel and Wulf,[9] and at the U.S. Bureau of Standards by Rose,[10] both in the 1930s.

In the 1940s, advances in instrumentation and the availability of commercial NIR instruments helped to bring about applications in the chemical and polymer industries. In 1949, Hibbard and Cleaves at the Lewis Flight Propulsion Laboratory in Cleveland, OH, studied hydrocarbon absorptions with the goal of analyzing octane and cetane numbers in reciprocating engine fuels, analyzing lubricating oils, and determining the structure of polymers.[11] In the late 1940s and early 1950s, Lauer and Rosenbaum of Sun Oil also worked with hydrocarbons in the second overtone region.[12] They mention studying natural and synthetic plastics including oriented nylon, polyvinyl chloride, and other polymers.

At about 1951, both Wilbur Kaye at Tennessee Eastman and Harry Willis in the Plastics Division of ICI in the U.K. began exploring NIR. Wilbur Kaye modified a Beckman DU UV instrument and worked on the development of the Beckman DK for NIR. Kaye thoroughly discussed the spectra of several compounds, including bromoform, chloroform, methylene chloride, benzene, methanol, and m-toluidine.[13] He cited potential application of the spectral region to the analysis of mixtures of organic compounds, including water in hydrocarbons and other solvents, alcohols in hydrocarbons, acids, amines, benzene, and olefins in hydrocarbons.

Harry Willis initially designed and built an instrument with a small CsI prism.[14] He worked with copolymer analyses, using thicker sheets of polymer than was possible with mid-infrared. Measurements were made quickly and directly.[15] He was able to estimate monomer content, determine molecular weight by end-group analysis, and measure copolymers such as butadiene/styrene.

Two other industrial chemists who worked with developing NIR applications in the 1950s were Robert Goddu of Hercules Powder Company (now Hercules Incorporated) and Kermit Whetsel of Tennessee Eastman. Goddu explored a number of different applications including the measurement of epoxide functionality and unsaturation in polymers and phenolics. His review chapter published in 1960 included many of these applications.[16] Whetsel also studied a number of applications including phenols, fuels, and polymers, and included those in a review article published in 1968.[17]

Following these and other industrial applications of NIR in the 1950s–1960s, the field became somewhat stagnant for laboratory analyses with the advent of NMR and other newer technologies that had greater distinguishing capabilities. Online process instruments, especially for water or moisture measurements, were an exception. These came into use in the 1940s and continued through today, but they were primarily filter instruments measuring only a few specific wavelengths.

With the rise of NIR reflection instrumentation and chemometric data treatments for rapid, direct control measurements, the interest in NIR for industrial applications became greater again. Process control applications, either rapid laboratory measurements or real-time online analyses, benefited the most from the new technologies. Older applications, such as hydroxyl number, polymer properties, water measurement, and fuel analyses have been revisited.

13.2 HISTORY OF NIR FOR FOOD AND AGRICULTURE

It is generally recognized that the history of the application of NIR to the food and agricultural industry began with Karl Norris at the USDA. In fact, Karl Norris' work was instrumental in the major renaissance of the NIR spectral region, especially with regard to its application to direct, solids-sampling measurements.

As Karl related in a recent review article,[18] the first work was done beginning in 1949 using visible- to low-wavelength NIR radiation to grade eggs. An unpublished spectrum taken in 1952 showed a water overtone at about 750 nm, but the peak was not found to be useful in predicting egg quality, and he did not pursue NIR for about 10 years in favor of visible spectrometry for color sorting.

In 1962, Norris, Hart, and Golumbic first published an NIR application, the determination of moisture in methanol extracts of seeds.[19] This was followed by transmission measurements of moisture in intact seeds with carbon tetrachloride used for reducing scattering losses.[20] It was the leap to diffuse reflection, however, that propelled NIR spectroscopy to wide application in agriculture and, eventually, many other industries as well. This shift required instrument modifications and the segment/gap derivative multiple linear regression approach to calibration that was pioneered in Norris's laboratory.

During the first stage in the development of NIR for food and agriculture, moisture analysis and nutrient determinations were driving forces. These measurements were important because the economic value of the products such as wheat is based on the dry weight and the protein content. Wet chemical methods for these analyses were costly and time-consuming, and not available at the points of sale. Early examples include the measurement of moisture in grain[21] and soybeans.[22]

In Canada, Phil Williams[23] of the Canadian Grain Commission purchased one of the first commercial instruments available and proceeded to initiate programs throughout his country to improve the measurement of moisture and protein in wheat. Similar efforts were launched in other large grain-producing countries such as Australia and Russia.

Applications in other agricultural products such as corn[24] and tobacco[25] followed. Forage analysis was another big driver of networked analyzer systems.[26] The technology then rapidly spread to processed foods such as chocolate, baked goods, meat and dairy products, and snack foods. In Japan, the bulk of the NIR applications are in food and agriculture.[27]

Agricultural products were primarily dependent upon a small number of "protein," fat, moisture, and carbohydrate wavelengths that were determined empirically through wavelength selection of large sample sets. In addition, the food and agricultural industries developed some nonchemical measurements, such as hardness of wheat, which was dependent upon particle size.[28] There were some limited efforts to perform spectral band assignments, but most of the calibrations were empirical. The complexity of natural products discourages very specific band assignments. For example, the measurement of "protein" involves grouping many types of protein molecules. Early wavelengths used to measure protein, oil, and water were the combination bands at 2180 nm for protein, 2305 nm for oil, and 1940 nm for water.[29] Tabulations of some general band assignments for foods and agricultural products can be found in the Osborne/Fearn and Williams/Norris books.

13.3 HISTORY OF NIR FOR PHARMACEUTICALS

NIR spectroscopy was used by pharmaceutical research groups at least since the late 1960s, initially to study hydrogen bonding of amines and amides or to perform quantitative analyses of these functional groups in solution. Examples include studies by S. Edward Krikorian at the University of Maryland,[30–33] Yumiko Tanaka and Katsunosuke Machida of Kyoto University,[34,35] and L. A. Strait and M. K. Hrenoff of the University of California,[36] all from pharmacy departments. In 1966, Sinsheimer and Keuhnelian[37] analyzed pharmaceutically active amine salts in pressed pellets. In 1967, Oi and Inaba determined phenacetin in preparations by NIR.[38] In a 1977 article, Zappala and Post reported using solution NIR to measure meprobamate in tablets, capsules, suspensions, and injectables.[39] This was an advantage over previous mid-infrared methodology because, in the NIR, the amine combination band at 1958 nm is separated from interference by hydroxyls and thus did not need to be chromatographically separated from an alcohol stabilizer.

The use of "modern" NIR in the pharmaceutical industry began soon after its use in the food and agricultural industries. Early applications included the measurement of moisture, particle size, the determination of composition, and identification of raw materials.

NIR moisture analyzers existed in industry long before the "modern" mode of multivariate NIR analysis, and the earliest applications of NIR in the pharmaceutical industry also began with moisture determinations. As examples, Beyer and Steffens[40] measured water in excipients directly, and Sinsheimer and Poswalk looked at water by extracting into acetontrile.[41]

Particle size was also a subject of early experiments. Although NIR spectroscopy alone cannot provide a particle size distribution the way other technologies such as image analysis, screening, and light scattering do, it is possible to estimate an average particle size by manipulating the underlying scattering curve. When coupled with low-angle light scattering, NIR has been shown to be quite accurate.[42]

Quantitative analysis of tablets and other dosage forms were also performed early in the current wave of NIR. In 1982, Rose et al. reported the direct analysis of meglumine and meglumine diatrizoate in injectable solutions, for example.[43] In 1986, Whitfield[44] developed a method to measure the amount of lincomycin in granulations for veterinary purposes. This was the first NIR method accepted by the FDA as a primary method.[45]

Whitfield's article also represents an example of the use of discriminant analysis in pharmaceutical methods. Probably the first work in this field was done by Rose in 1982,[46] when he showed that a number of structurally similar drugs could be identified. This was followed by the application of Mahalanobis distances to pharmaceutical raw materials.[47]

A review of the uses of NIR in pharmaceuticals was done in 1987,[48] and a chapter written in 1992 and revised in 2001.[49] The field continues to expand.

REFERENCES

1. Herschel, W., Experiments in the refrangibility of the invisible rays of the sun, *Phil. Trans. R. Soc. London*, 90(XIV), 284–292, 1800.
2. Abney, R.E. and Festing, R.E., On the influence of the atomic grouping in the molecules of organic bodies on their absorption in the infra-red region of the spectrum, *Phil. Trans. R. Soc. London*, 172(887–918), 1881.
3. Donath, B., *Ann. Phys.*, 58(3), 609, 1896.
4. Puccianti, L., *Nuovo Cim.*, 11, 141, 1900; *Phys. Zeit.*, 1, 48, 1899; *Phys. Zeit.*, 494, 1900.
5. Coblentz, W.W., *Astrophys. J.*, 20, 1904.
6. Coblentz, W.W., *Investigations of Infrared Spectra*. Part i: Absorption spectra part ii: Infrared absorption spectra part ii: Infrared emission spectra, Carnegie Institution of Washington (Pub. No. 35) Washington, D.C., 1905, pp. 1–330.
7. Whetsel, K.B., The first fifty years of near-infrared spectroscopy in America, *NIR News*, 2(3), 4–5, 1991.

8. Ellis, J.W., The near-infrared absorption spectra of some organic liquids, *Phys. Rev.*, 23(1), 48–62, 1924.
9. Liddel, U. and Wulf, O.R., *J. Am. Chem. Soc.*, 55, 1933.
10. Rose, F.W., Jr., Quantitative analysis, with respect to the component structural groups, of the infrared (1 to 2 u) molal absorptive indices of 55 hydrocarbons, *J. Res. Natl. Bur. Stand.*, 20,129–157, 1938.
11. Hibbard, R.R. and Cleaves, A.P., Carbon-hydrogen groups in hydrocarbons. Characterization by 1.10 to 1.25-micron infrared absorption, *Anal. Chem.*, 21(4), 486–492, 1949.
12. Lauer, J.L. and Rosenbaum, E.J., Near-infrared absorption spectrometry, *Appl. Spectrosc.*, 6(5), 29–40, 46, 1952.
13. Kaye, W., Near-infrared spectroscopy i. Spectral identification and analytical applications, *Spectrochim. Acta*, 6, 257–287, 1954.
14. Miller, R., Professor Harry Willis and the history of NIR spectroscopy, *NIR News*, 2(4), 12–13, 1991.
15. Miller, R.G.J. and Willis, H.A., Quantitative analysis in the 2-u region applied to synthetic polymers, *J. Appl. Chem.*, 6, 385–391, 1956.
16. Goddu, R.F., Near-infrared spectrophotometry, *Adv. Anal. Chem. Instrum.*, 1, 347, 1960.
17. Whetsel, K.B., Near-infrared spectrophotometry, *Appl. Spectrosc. Rev.*, 2, 1–67, 1968.
18. Norris, K., Early history of near-infrared for agricultural applications, *NIR News*, 12–13, 1992.
19. Hart, J.R. and Norris, K.H., Golumbic, determination of the moisture content of seeds by near-infrared spectrophotometry of their methanol extracts, *Cereal Chem.*, 39, 94–99, 1962.
20. Norris, K.H. and Hart, J.R., Direct spectrophotometric determination of moisture content of grain and seeds, in *Proc. 1963 Int. Symp. Humidity Moisture*, Vol. 4, Reinhold, New York, 1965, pp. 19–25.
21. Massie, D.R. and Norris, K.H., The spectral reflectance and transmittance properties of grain in the visible and near-infrared, *Trans. Am. Soc. Agric. Eng.*, 8(4), 598–600, 1964.
22. Ben-Gera, I. and Norris, K.H., Determination of moisture content in soybeans by direct spectrophotometry, *Isr. J. Agric. Res.*, 18(3), 125–132, 1968.
23. Williams, P.C., Application of near-infrared reflectance spectroscopy to analysis of cereal grains and oilseeds, *Cereal Chem.*, 52(4), 561–576, 1975.
24. Finney, E.E., Jr. and Norris, K.H., Determination of moisture in corn kernels by near-infrared-transmittance measurements, *Trans. Am. Soc. Agric. Eng.*, 21(3), 581–584, 1978.
25. McClure, W.F., Norris, K.H., and Weeks, W.W., Rapid spectrophotometric analysis of the chemical composition of tobacco. Part 1: Total reducing sugars, *Beitr. Tabakforsch.*, 9(1), 13–18, 1977.
26. Shenk, J.S., Westerhaus, M.O., and Hoover, M.R., Analysis of forages by infrared reflectance, *J. Dairy Sci.*, 62, 807–812, 1979.
27. Iwamoto, M., Kawano, S., and Ozaki, Y., An overview of research and development of near-infrared spectroscopy in Japan, *J. Near-Infrared Spectrosc.*, 179–189, 1995.
28. Williams, P.C., Screening wheat for protein and hardness by near-infrared reflectance spectroscopy, *Cereal Chem.*, 56(3), 169–172, 1979.
29. Norris, K.H., Near-infrared reflectance spectroscopy — the present and future. Cereals' 78: better nutrition for the world's millions, *6th Int. Cereal Bread Congr.*, pub. AACC, 1978, pp. 245–252.
30. Krikorian, S.E. and Mahpour, M., The identification and origin of n-h overtone and combination bands in the near-infrared spectra of simple 1 and 2 amides, *Spectrochim. Acta*, 29A, 1233, 1973.
31. Krikorian, S.E. and Mahpour, M., The origin of amide group overtone and combination bands in the near-infrared spectra of cis-configured lactams, *Spectrochim. Acta*, 32A, 1447–1453, 1976.
32. Krikorian, S.E., The influence of hydrogren-bonded association on the character of the near-infrared spectra of simple cis- and trans-configured secondary amides, *Spectrochim. Acta*, 37A(9), 745–751, 1981.
33. Krikorian, S.E., Determination of dimerization constants of cis- and trans-configured secondary amides using near-infrared spectrophotometry, *J. Phys. Chem.*, 86, 1875–1881, 1982.
34. Tanaka, Y. and Machida, K., Anharmonicity of NH_2-stretching vibrations of substituted anilines, *J. Mol. Spectrosc.*, 55, 435–444, 1975.
35. Tanaka, Y. and Machida, K., Near-infrared spectra of formamide and its anharmonic potential, *J. Mol. Spectrosc.*, 63, 306–316, 1976.
36. Strait, L.A. and Hrenoff, M.K., Near-infrared spectrophotometric study of hydrogen bonding in primary aliphatic amines, *Spectrosc. Lett.*, 8(4), 165–174, 1975.
37. Sinsheimer, J.E. and Keuhnelian, A.M., Near-infrared spectroscopy of amine salts, *J. Pharm. Sci.*, 55(11), 1240–1244, 1966.

38. Oi, N. and Inaba, E., Analyses of drugs and chemicals by infrared absorption spectroscopy. 8. Determination of allylisopropylacetureide and phenacetin in pharmaceutical preparations by near-infrared absorption spectroscopy, *Yakugaku Zasshi*, 87(3), 213–215, 1967.

39. Zappala, A.F. and Post, A., Rapid near-infrared spectrophotometric determination of Meprobamate in pharmaceutical preparations, *J. Pharm. Sci.*, 66(2), 292–293, 1977.

40. Beyer, J. and Steffens, K., Calibration models for determination of water content in pharmaceutical excipients using near-infrared spectroscopy (nirs), *Pharm. Ind.* (Germany), 65(2), 186–192, 1975.

41. Sinsheimer, J.E. and Poswalk, N.M., Pharmaceutical applications of the near-infrared determination of water, *J. Pharm. Sci.*, 57(11), 2007–2010, 1968.

42. Ciurczak, E.W., Torlini, R.P., and Demkowicz, M.P., Determination of particle size of pharmaceutical raw materials using near-infrared reflectance spectroscopy, *Spectroscopy*, 1(7), 36–40, 1986.

43. Rose, J.J. and Prusik, T. et al., Near-infrared mult-component analysis of parenteral products using the infraanalyzer 400, *J. Parenter. Sci. Technol.*, 36(2), 71–78, 1982.

44. Whitfield, R.G., *Pharm. Manuf.*, 31–40, 1986.

45. Ciurczak, E.W., NIR analysis of pharmaceuticals, in *Handbook of Near-Infrared Analysis*, 1st ed., Burns, D.A. and Ciurczak, E.W. (Eds.), Marcel Dekker, New York, 1992, pp. 549–563.

46. Rose, J.R., *Proc. Annu. Symp. NIRA*, Technicon, Tarrytown, New York, 1982.

47. Ciurczak, E.W., *Proc. Annu. Symp. NIRA*, Technicon, Tarrytown, New York, 1984.

48. Ciurczak, E.W., Uses of near-infrared spectroscopy in pharmaceutical analysis, *Appl. Spectrosc. Rev.*, 23(1,2), 147–163, 1987.

49. Ciurczak, E.W. and Drennen, J., NIR analysis of pharmaceuticals, in *Handbook of Near-Infrared Analysis*, 2nd ed., Burns, D.A. and Ciurczak, E.W. (Eds.), Marcel Dekker, New York, 2001, pp. 609–632.

Appendix 1: NIR Absorption Band Charts — An Overview

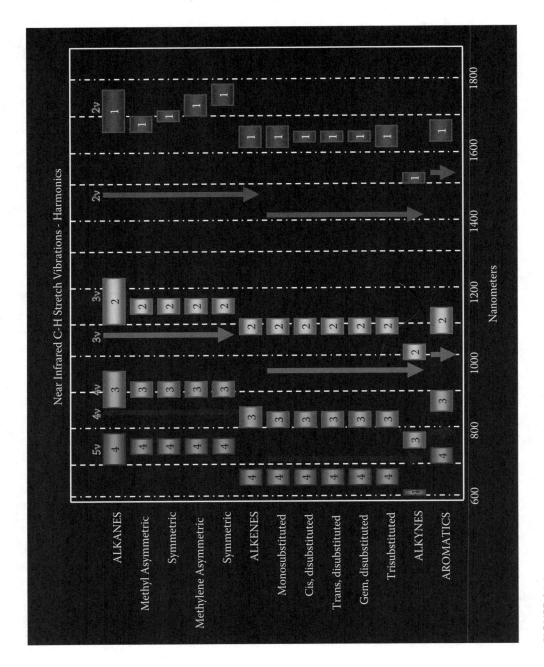

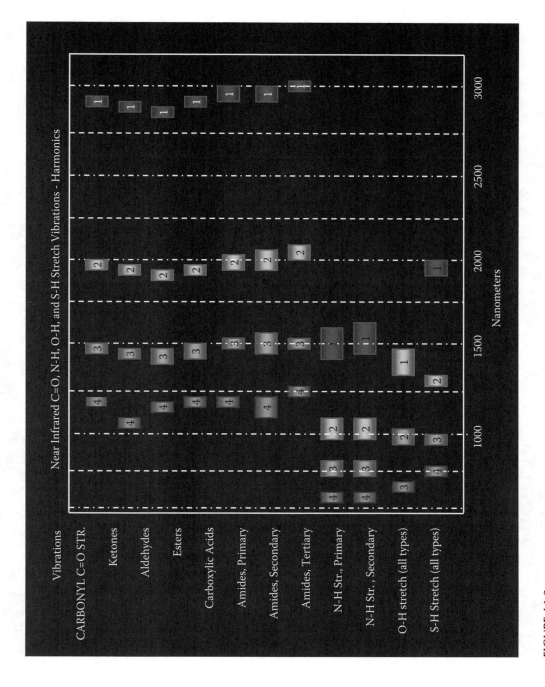

FIGURE A1.2

Appendix 2a: Spectra–Structure Correlations—Labeled Spectra from 10,500 cm^{-1} to 6300 cm^{-1} (952 nm to 1587 nm)

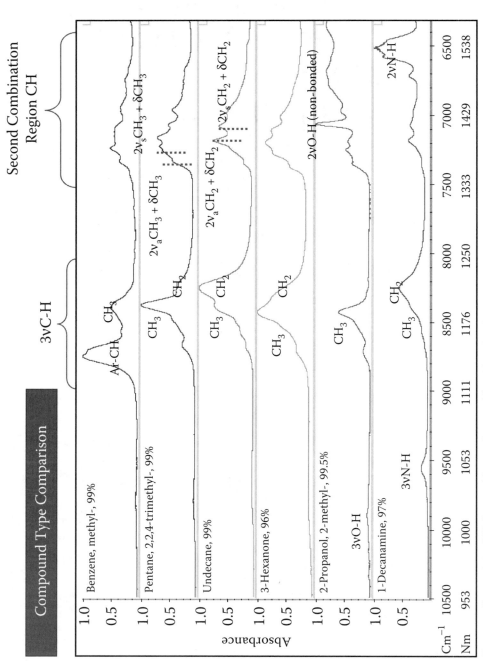

Spectra used by permission "NIR Spectra of Organic Compounds," © Wiley-VCH, ISBN 3-527-31630-2.

FIGURE A2a.1

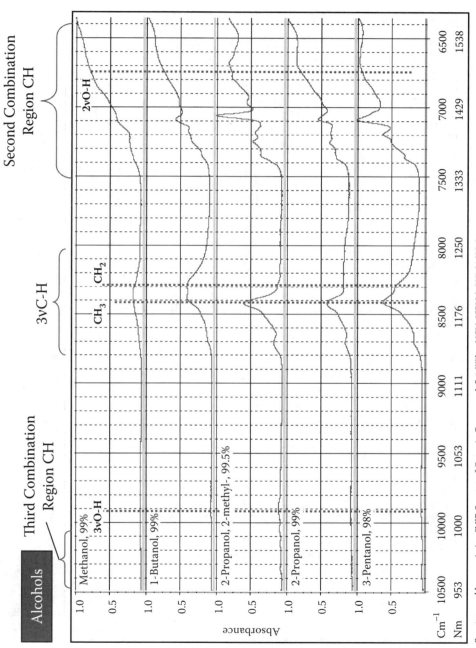

Spectra used by permission "NIR Spectra of Organic Compounds," © Wiley-VCH, ISBN 3-527-31630-2.

FIGURE A2a.2

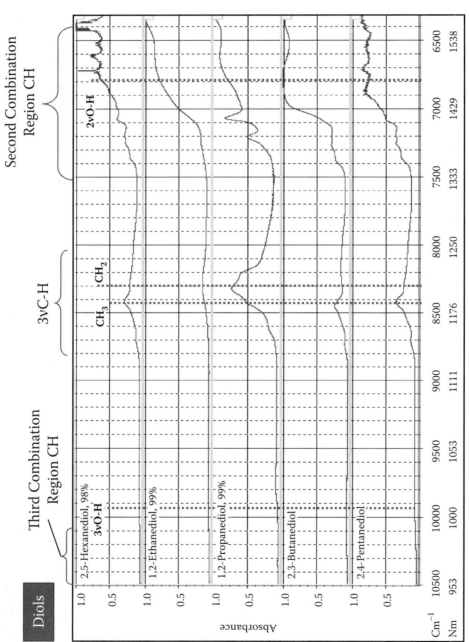

Spectra used by permission "NIR Spectra of Organic Compounds," © Wiley-VCH, ISBN 3-527-31630-2.

FIGURE A2a.3

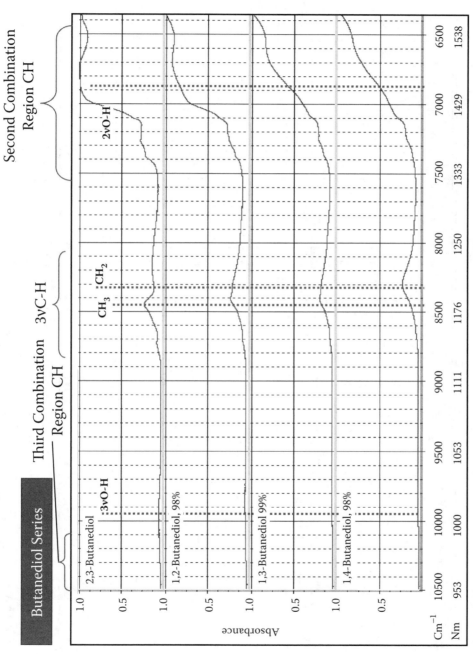

Second Combination Region CH

Third Combination   3vC-H
Region CH

Butanediol Series

2,3-Butanediol

3vO-H

2vO-H

CH₃ :CH₂

1,2-Butanediol, 98%

1,3-Butanediol 99%

1,4-Butanediol, 98%

Absorbance

| Cm⁻¹ | 10500 | 10000 | 9500 | 9000 | 8500 | 8000 | 7500 | 7000 | 6500 |
| Nm | 953 | 1000 | 1053 | 1111 | 1176 | 1250 | 1333 | 1429 | 1538 |

Spectra used by permission "NIR Spectra of Organic Compounds," © Wiley-VCH, ISBN 3-527-31630-2.

FIGURE A2a.4

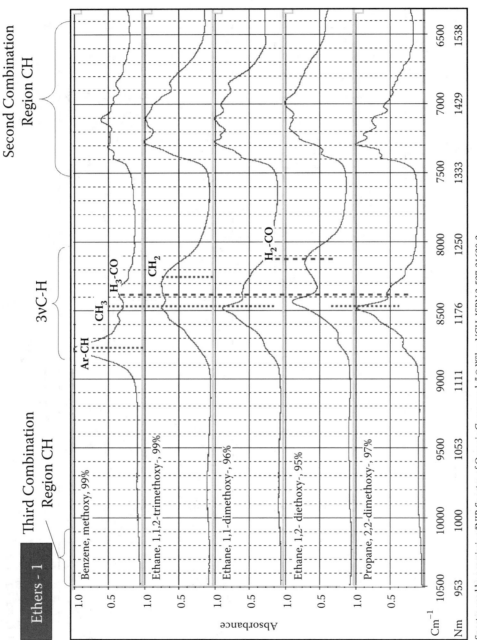

Spectra used by permission "NIR Spectra of Organic Compounds," © Wiley-VCH, ISBN 3-527-31630-2.

FIGURE A2a.5

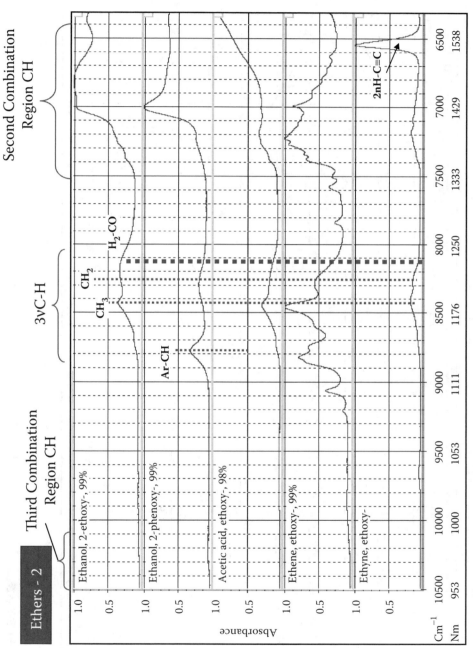

Spectra used by permission "NIR Spectra of Organic Compounds," © Wiley-VCH, ISBN 3-527-31630-2.

FIGURE A2a.6

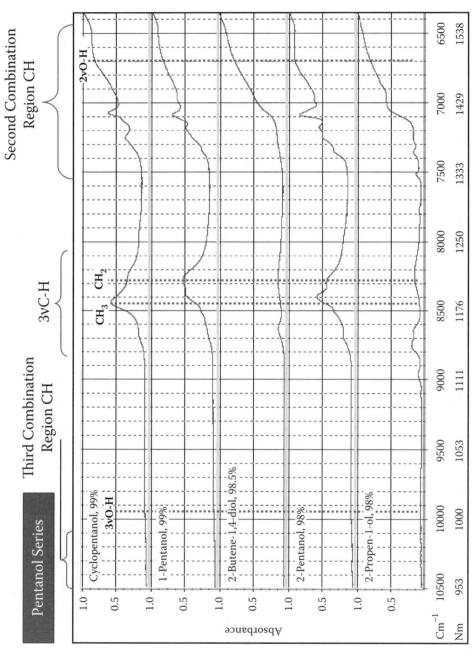

Spectra used by permission "NIR Spectra of Organic Compounds," © Wiley-VCH, ISBN 3-527-31630-2.

FIGURE A2a.7

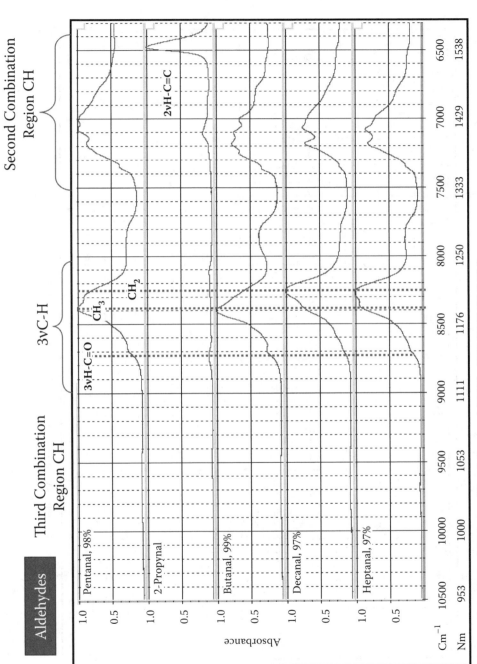

Aldehydes

Third Combination Region CH

3νC-H

Second Combination Region CH

Pentanal, 98%

3νH-C=O

CH₃

CH₂

2-Propynal

2νH-C≡C

Butanal, 99%

Decanal, 97%

Heptanal, 97%

Absorbance

| Cm⁻¹ | 10500 | 10000 | 9500 | 9000 | 8500 | 8000 | 7500 | 7000 | 6500 |
| Nm | 953 | 1000 | 1053 | 1111 | 1176 | 1250 | 1333 | 1429 | 1538 |

Spectra used by permission "NIR Spectra of Organic Compounds," © Wiley-VCH, ISBN 3-527-31630-2.

FIGURE A2a.8

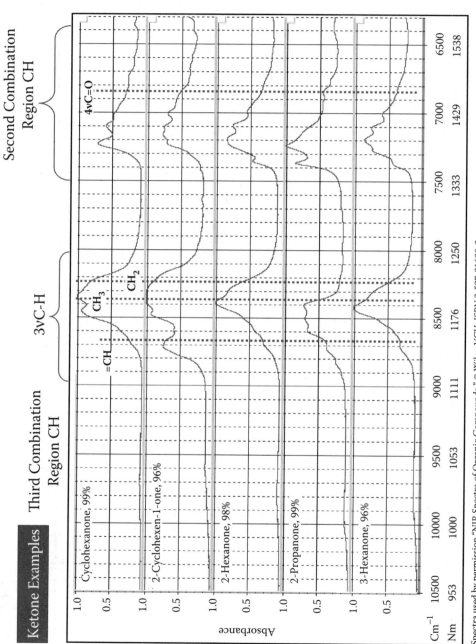

FIGURE A2a.9

Spectra used by permission "NIR Spectra of Organic Compounds," © Wiley-VCH, ISBN 3-527-31630-2.

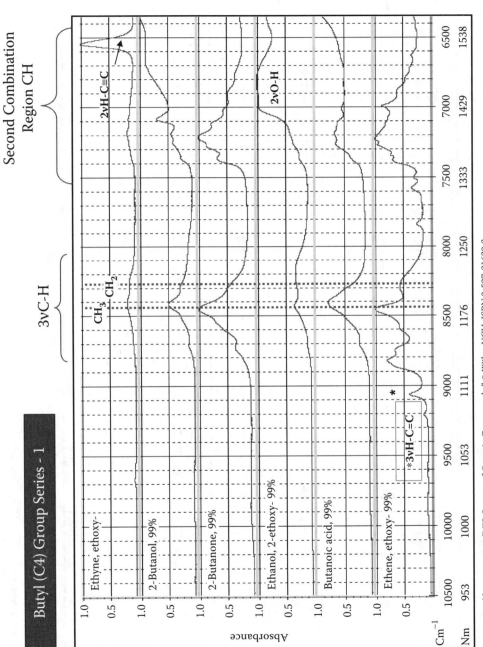

Butyl (C4) Group Series - 1

FIGURE A2a.10

Spectra used by permission "NIR Spectra of Organic Compounds," © Wiley-VCH, ISBN 3-527-31630-2.

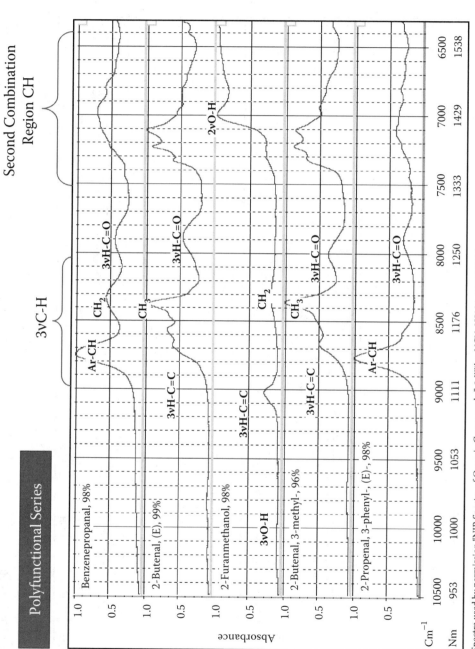

Spectra used by permission "NIR Spectra of Organic Compounds," © Wiley-VCH, ISBN 3-527-31630-2.

FIGURE A2a.11

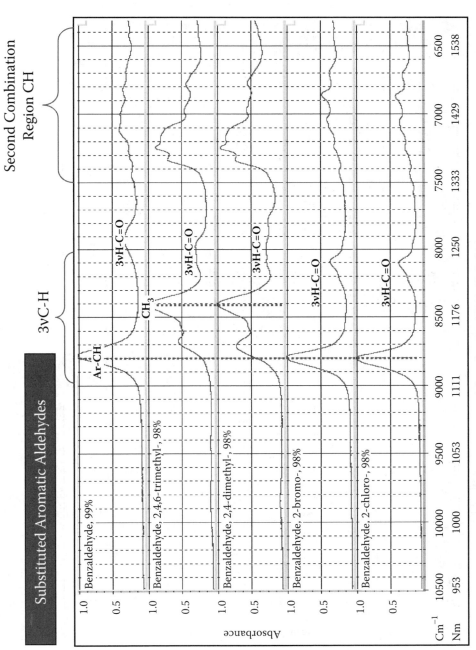

Second Combination Region CH

3vC–H

Substituted Aromatic Aldehydes

Ar-CH

CH₃

3vH-C=O

3vH-C=O

3vH-C=O

3vH-C=O

3vH-C=O

Benzaldehyde, 99%

Benzaldehyde, 2,4,6–trimethyl-, 98%

Benzaldehyde, 2,4–dimethyl-, 98%

Benzaldehyde, 2-bromo-, 98%

Benzaldehyde, 2-chloro-, 98%

Absorbance

| Cm⁻¹ | 10500 | 10000 | 9500 | 9000 | 8500 | 8000 | 7500 | 7000 | 6500 |
| Nm | 953 | 1000 | 1053 | 1111 | 1176 | 1250 | 1333 | 1429 | 1538 |

Spectra used by permission "NIR Spectra of Organic Compounds," © Wiley-VCH, ISBN 3-527-31630-2.

FIGURE A2a.12

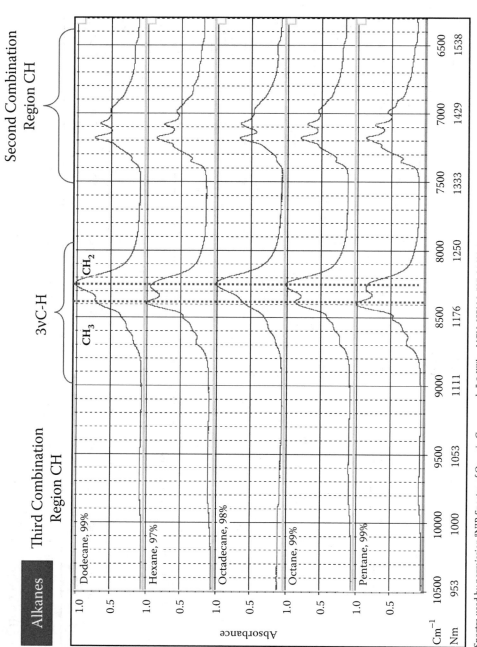

Spectra used by permission "NIR Spectra of Organic Compounds," © Wiley-VCH, ISBN 3-527-31630-2.

FIGURE A2a.13

Butyl (C4) Group Series

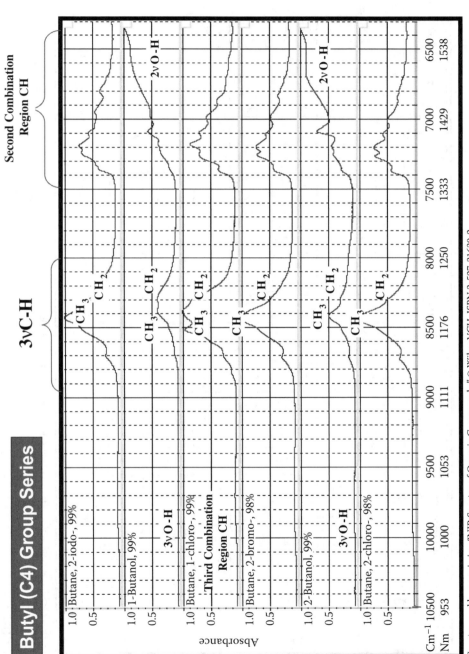

Spectra used by permission "NIR Spectra of Organic Compounds," © Wiley-VCH, ISBN 3-527-31630-2.

FIGURE A2a.14

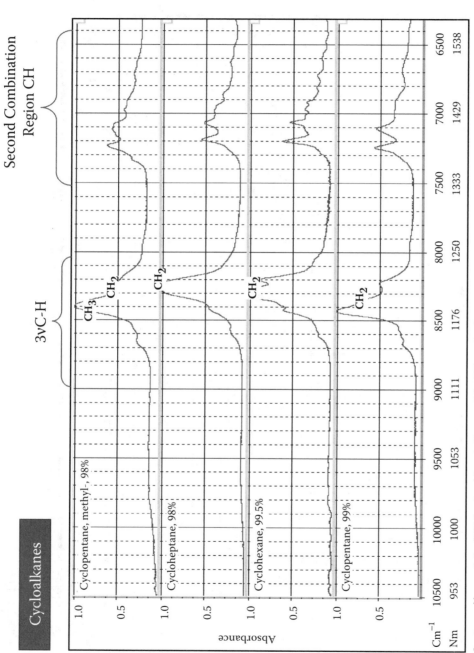

Spectra used by permission "NIR Spectra of Organic Compounds," © Wiley-VCH, ISBN 3-527-31630-2.

FIGURE A2a.15

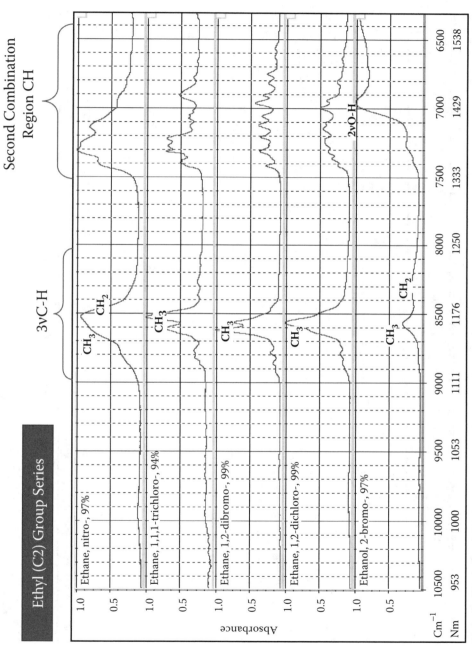

Ethyl (C2) Group Series

Second Combination Region CH

3νC-H

Ethane, nitro-, 97%

Ethane, 1,1,1-trichloro-, 94%

Ethane, 1,2-dibromo-, 99%

Ethane, 1,2-dichloro-, 99%

Ethanol, 2-bromo-, 97%

CH₃

CH₂

2νO-H

Absorbance

| Cm⁻¹ | 10500 | 10000 | 9500 | 9000 | 8500 | 8000 | 7500 | 7000 | 6500 |
| Nm | 953 | 1000 | 1053 | 1111 | 1176 | 1250 | 1333 | 1429 | 1538 |

Spectra used by permission "NIR Spectra of Organic Compounds," © Wiley-VCH, ISBN 3-527-31630-2.

FIGURE A2a.16

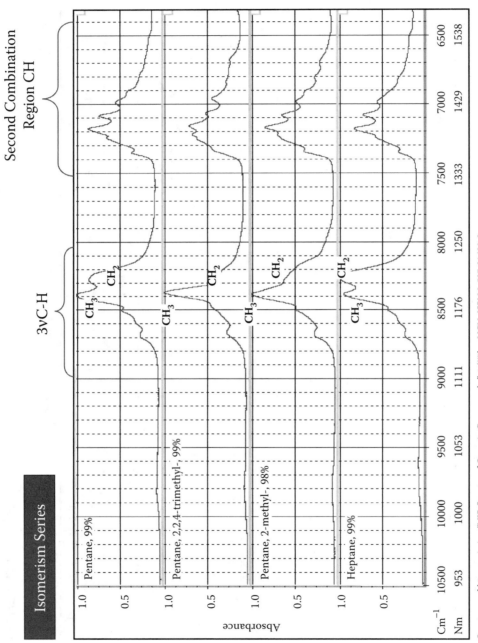

Spectra used by permission "NIR Spectra of Organic Compounds," © Wiley-VCH, ISBN 3-527-31630-2.

FIGURE A2a.17

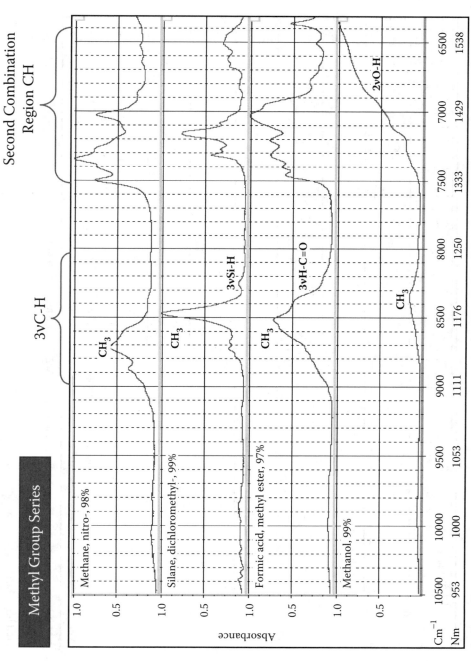

Second Combination Region CH

3vC-H

Methyl Group Series

Methane, nitro-, 98%

CH₃

Silane, dichloromethyl-, 99%

CH₃

3vSi-H

Formic acid, methyl ester, 97%

CH₃

3vH-C=O

Methanol, 99%

CH₃

2vO-H

Absorbance

| Cm⁻¹ | 10500 | 10000 | 9500 | 9000 | 8500 | 8000 | 7500 | 7000 | 6500 |
| Nm | 953 | 1000 | 1053 | 1111 | 1176 | 1250 | 1333 | 1429 | 1538 |

Spectra used by permission "NIR Spectra of Organic Compounds," © Wiley-VCH, ISBN 3-527-31630-2.

FIGURE A2a.18

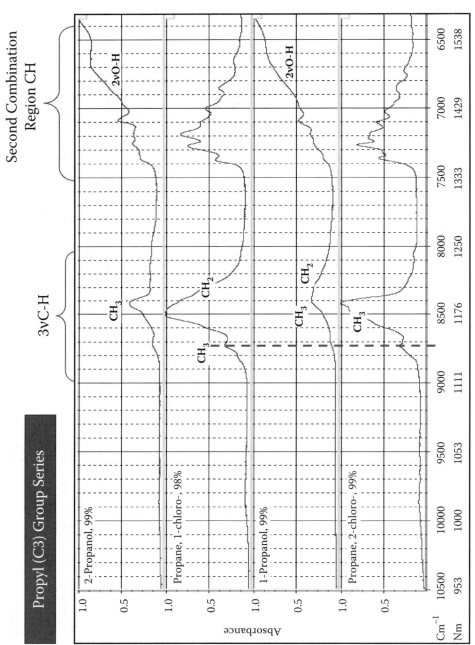

Spectra used by permission "NIR Spectra of Organic Compounds," © Wiley-VCH, ISBN 3-527-31630-2.

FIGURE A2a.19

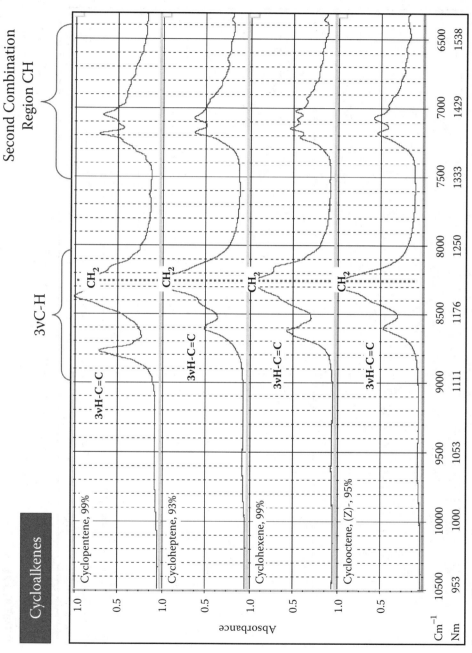

Second Combination Region CH

3νC-H

Cycloalkenes

Cyclopentene, 99%

Cycloheptene, 93%

Cyclohexene, 99%

Cyclooctene, (Z)-, 95%

3νH-C=C

3νH-C=C

3νH-C=C

3νH-C=C

CH₂

CH₂

CH₂

CH₂

Absorbance

| Cm⁻¹ | 10500 | 10000 | 9500 | 9000 | 8500 | 8000 | 7500 | 7000 | 6500 |
| Nm | 953 | 1000 | 1053 | 1111 | 1176 | 1250 | 1333 | 1429 | 1538 |

Spectra used by permission "NIR Spectra of Organic Compounds," © Wiley-VCH, ISBN 3-527-31630-2.

FIGURE A2a.20

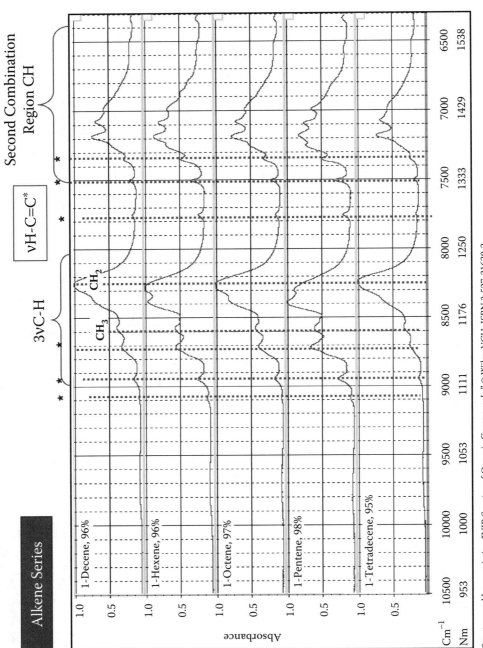

Spectra used by permission "NIR Spectra of Organic Compounds," © Wiley-VCH, ISBN 3-527-31630-2.

FIGURE A2a.21

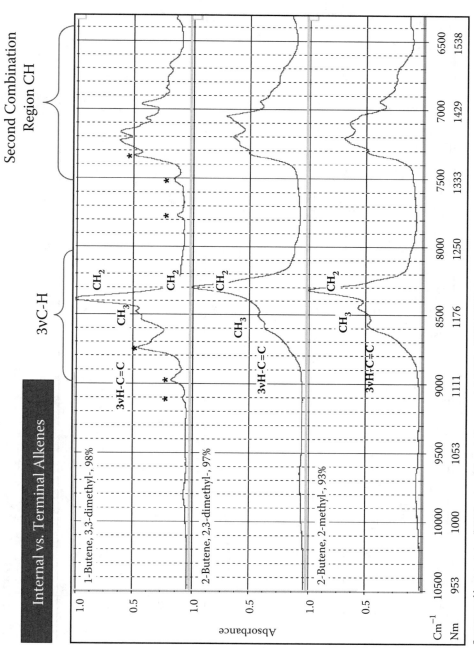

Internal vs. Terminal Alkenes

Second Combination Region CH

3νC–H

1-Butene, 3,3-dimethyl-, 98%

2-Butene, 2,3-dimethyl-, 97%

2-Butene, 2-methyl-, 93%

CH₂

CH₃

3νH–C=C

CH₂

CH₃

3νH–C=C

CH₂

CH₃

3νH–C=C

Spectra used by permission "NIR Spectra of Organic Compounds," © Wiley-VCH, ISBN 3-527-31630-2.

FIGURE A2a.22

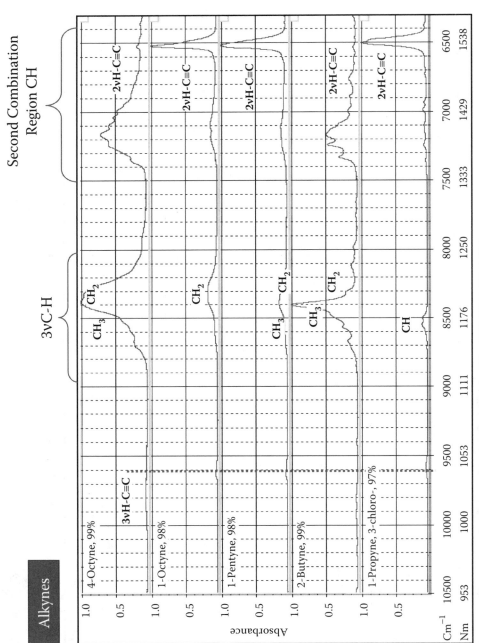

Spectra used by permission "NIR Spectra of Organic Compounds," © Wiley-VCH, ISBN 3-527-31630-2.

FIGURE A2a.23

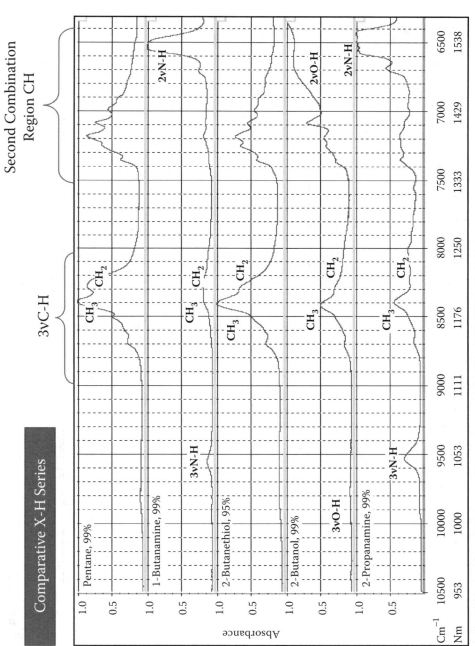

Spectra used by permission "NIR Spectra of Organic Compounds," © Wiley-VCH, ISBN 3-527-31630-2.

FIGURE A2a.24

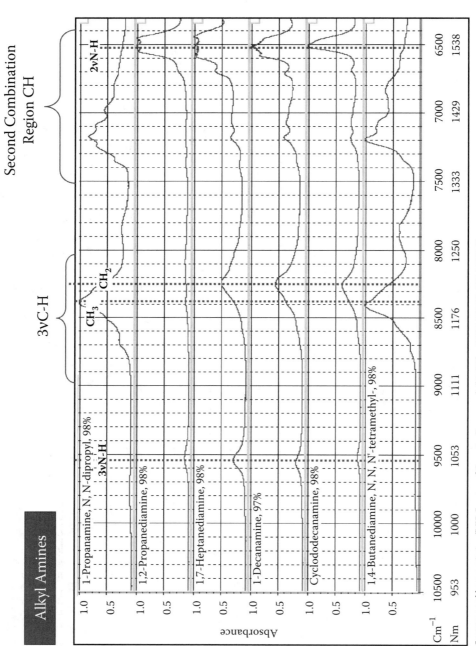

FIGURE A2a.25

Spectra used by permission "NIR Spectra of Organic Compounds," © Wiley-VCH, ISBN 3-527-31630-2.

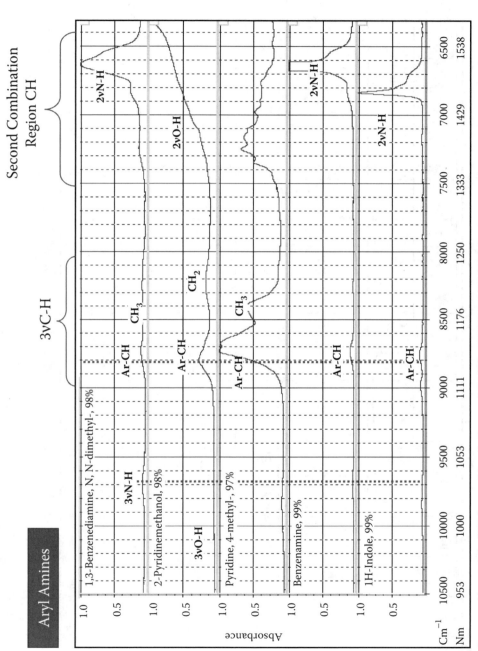

Spectra used by permission "NIR Spectra of Organic Compounds," © Wiley-VCH, ISBN 3-527-31630-2.

FIGURE A2a.26

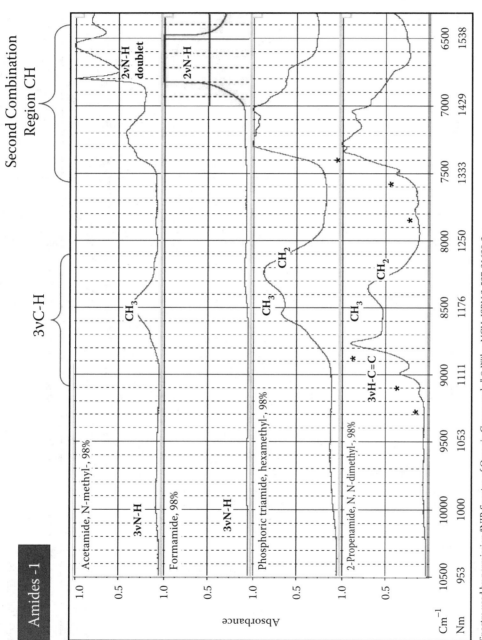

Spectra used by permission "NIR Spectra of Organic Compounds," © Wiley-VCH, ISBN 3-527-31630-2.

FIGURE A2a.27

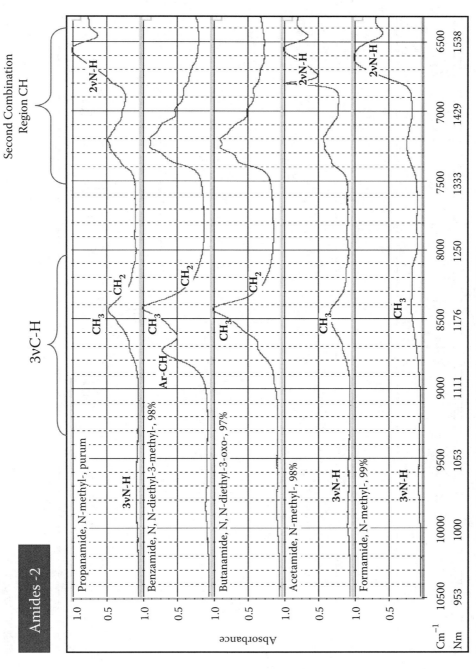

Amides -2

Second Combination Region CH

3νC-H

Propanamide, N-methyl-, purum

2νN-H

CH₃

CH₂

3νN-H

Benzamide, N, N-diethyl-3-methyl-, 98%

CH₃

CH₂

Ar-CH

Butanamide, N, N-diethyl-3-oxo-, 97%

CH₃

CH₂

2νN-H

3νN-H

Acetamide, N-methyl-, 98%

CH₃

2νN-H

3νN-H

Formamide, N-methyl-, 99%

CH₃

2νN-H

3νN-H

Absorbance

| Cm⁻¹ | 10500 | 10000 | 9500 | 9000 | 8500 | 8000 | 7500 | 7000 | 6500 |
| Nm | 953 | 1000 | 1053 | 1111 | 1176 | 1250 | 1333 | 1429 | 1538 |

Spectra used by permission "NIR Spectra of Organic Compounds," © Wiley-VCH, ISBN 3-527-31630-2.

FIGURE A2a.28

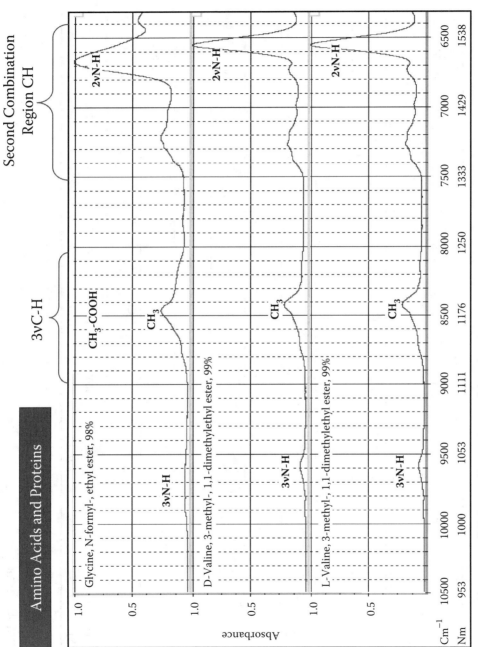

Spectra used by permission "NIR Spectra of Organic Compounds," © Wiley-VCH, ISBN 3-527-31630-2.

FIGURE A2a.29

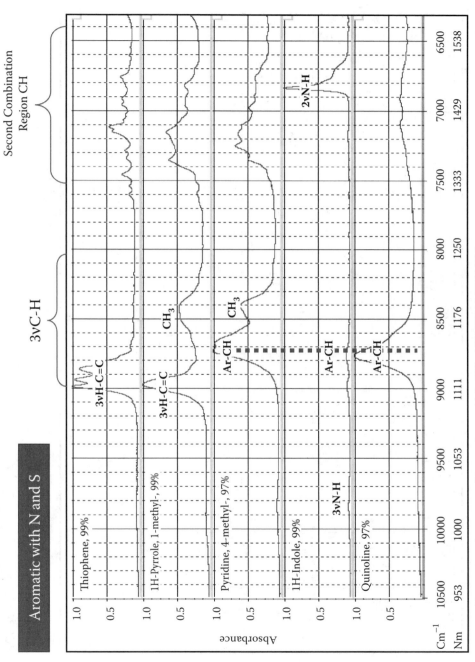

Spectra used by permission "NIR Spectra of Organic Compounds," © Wiley-VCH, ISBN 3-527-31630-2.

FIGURE A2a.30

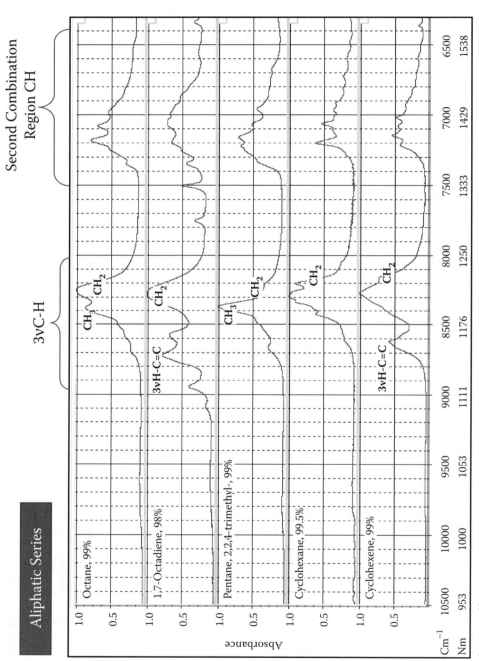

Spectra used by permission "NIR Spectra of Organic Compounds," © Wiley-VCH, ISBN 3-527-31630-2.

FIGURE A2a.31

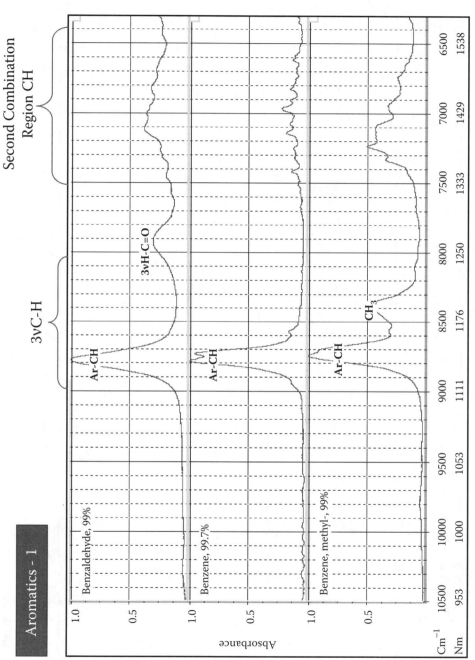

Aromatics – 1

Second Combination Region CH

3νC-H

Benzaldehyde, 99%

Ar-CH

3νH-C=O

Benzene, 99.7%

Ar-CH

Benzene, methyl-, 99%

Ar-CH

CH₃

Absorbance

| Cm⁻¹ | 10500 | 10000 | 9500 | 9000 | 8500 | 8000 | 7500 | 7000 | 6500 |
| Nm | 953 | 1000 | 1053 | 1111 | 1176 | 1250 | 1333 | 1429 | 1538 |

Spectra used by permission "NIR Spectra of Organic Compounds," © Wiley-VCH, ISBN 3-527-31630-2.

FIGURE A2a.32

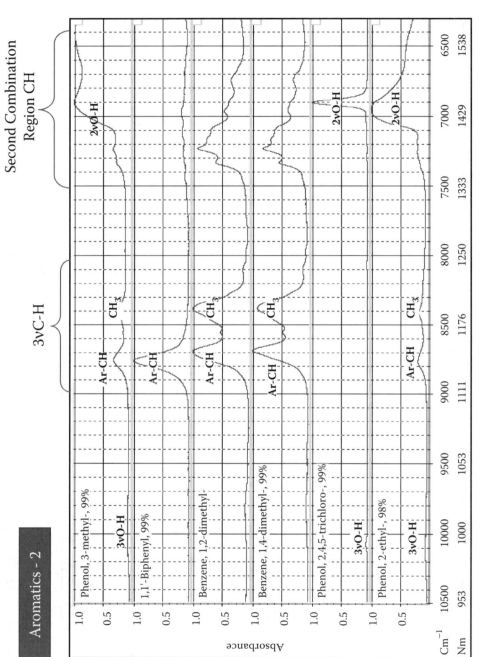

Aromatics – 2

Spectra used by permission "NIR Spectra of Organic Compounds," © Wiley-VCH, ISBN 3-527-31630-2.

FIGURE A2a.33

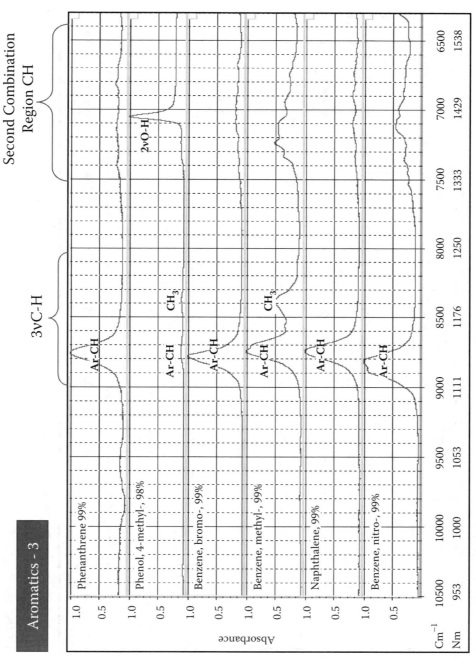

Second Combination Region CH

3vC-H

Aromatics - 3

Phenanthrene 99%

Ar-CH

Phenol, 4-methyl-, 98%

2vO-H

Ar-CH

CH₃

Benzene, bromo-, 99%

Ar-CH

Benzene, methyl-, 99%

CH₃

Ar-CH

Naphthalene, 99%

Ar-CH

Benzene, nitro-, 99%

Ar-CH

Absorbance

| Cm⁻¹ | 10500 | 10000 | 9500 | 9000 | 8500 | 8000 | 7500 | 7000 | 6500 |
| Nm | 953 | 1000 | 1053 | 1111 | 1176 | 1250 | 1333 | 1429 | 1538 |

1.0 / 0.5

Spectra used by permission "NIR Spectra of Organic Compounds," © Wiley-VCH, ISBN 3-527-31630-2.

FIGURE A2a.34

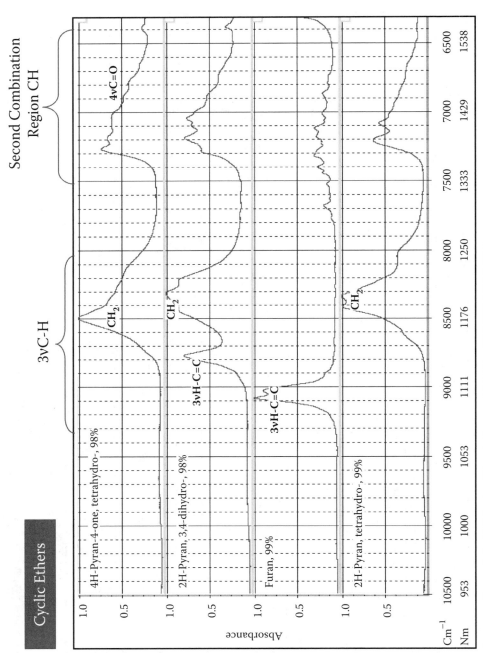

Spectra used by permission "NIR Spectra of Organic Compounds," © Wiley-VCH, ISBN 3-527-31630-2.

FIGURE A2a.35

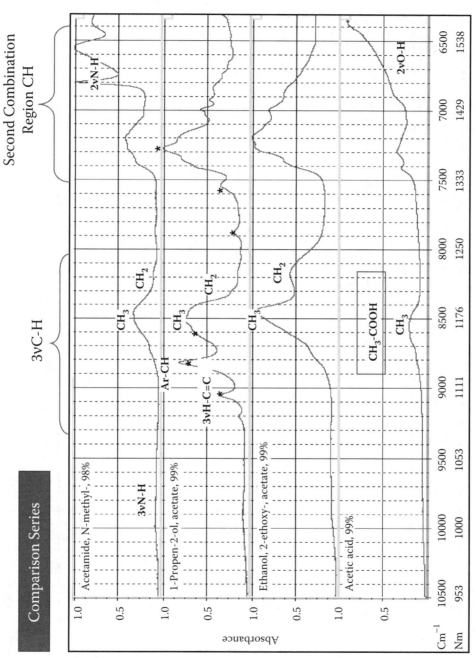

Spectra used by permission "NIR Spectra of Organic Compounds," © Wiley-VCH, ISBN 3-527-31630-2.

FIGURE A2a.36

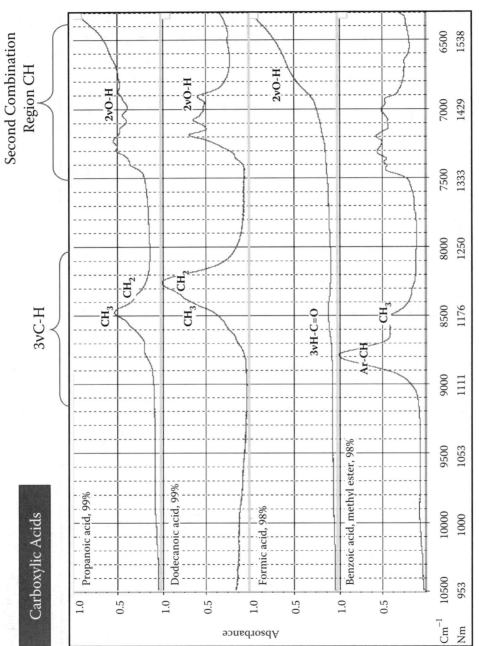

Carboxylic Acids

Spectra used by permission "NIR Spectra of Organic Compounds," © Wiley-VCH, ISBN 3-527-31630-2.

FIGURE A2a.37

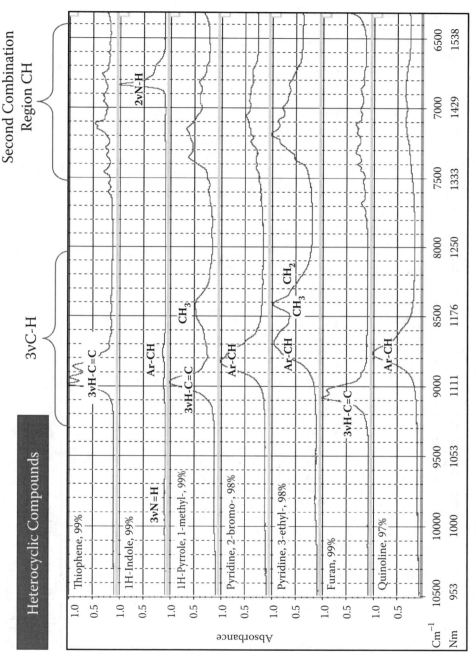

Second Combination Region CH

3νC–H

Heterocyclic Compounds

Thiophene, 99%

3νH–C=C

1H-Indole, 99%

3νN–H

2νN–H

Ar-CH

1H-Pyrrole, 1-methyl-, 99%

3νH–C=C

CH₃

Pyridine, 2-bromo-, 98%

Ar-CH

Pyridine, 3-ethyl-, 98%

Ar-CH

CH₃

CH₂

Furan, 99%

3νH–C=C

Quinoline, 97%

Ar-CH

Absorbance

| Cm⁻¹ | 10500 | 10000 | 9500 | 9000 | 8500 | 8000 | 7500 | 7000 | 6500 |
| Nm | 953 | 1000 | 1053 | 1111 | 1176 | 1250 | 1333 | 1429 | 1538 |

Spectra used by permission "NIR Spectra of Organic Compounds," © Wiley-VCH, ISBN 3-527-31630-2.

FIGURE A2a.38

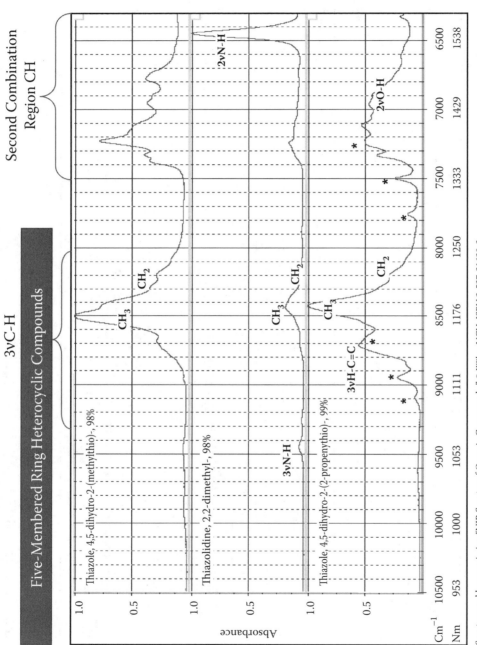

Spectra used by permission "NIR Spectra of Organic Compounds," © Wiley-VCH, ISBN 3-527-31630-2.

FIGURE A2a.39

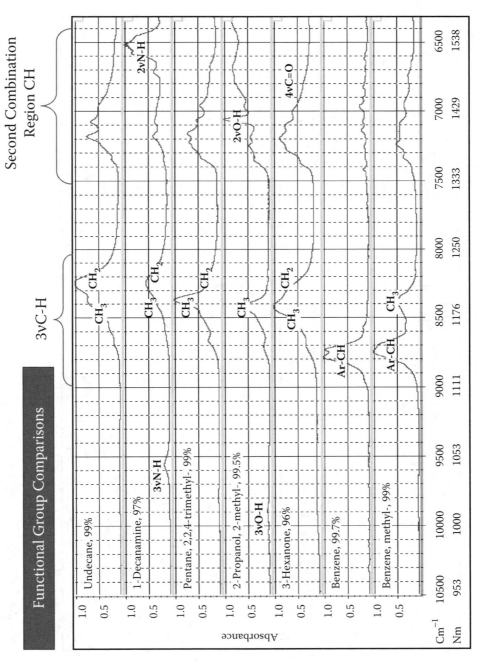

Functional Group Comparisons

Second Combination Region CH

3vC-H

Absorbance

Spectra used by permission "NIR Spectra of Organic Compounds," © Wiley-VCH, ISBN 3-527-31630-2.

FIGURE A2a.40

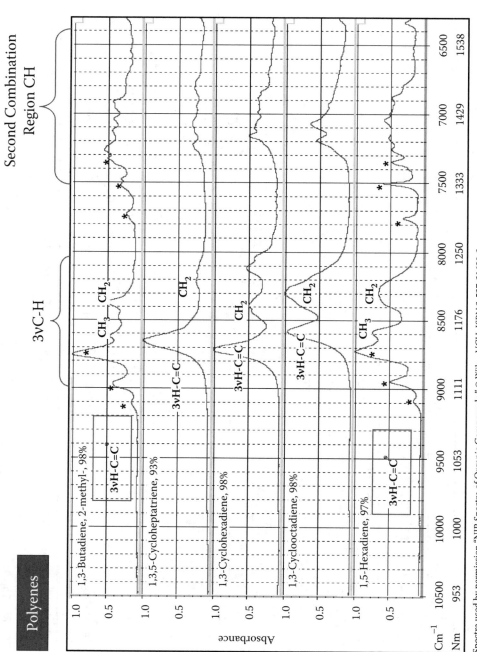

Spectra used by permission "NIR Spectra of Organic Compounds," © Wiley-VCH, ISBN 3-527-31630-2.

FIGURE A2a.41

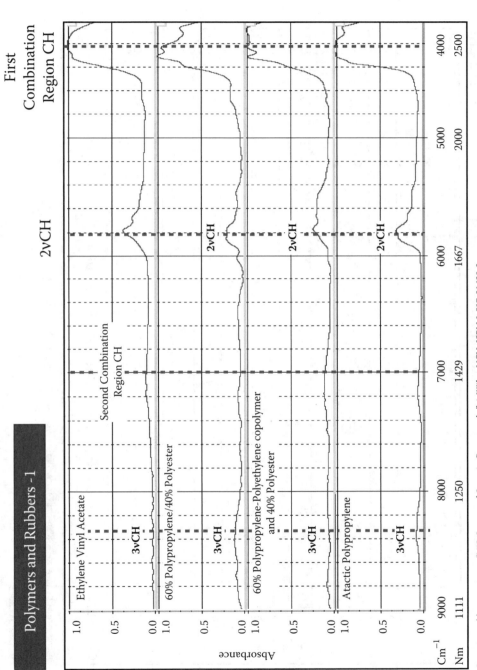

Polymers and Rubbers -1

First Combination Region CH

2vCH

Second Combination Region CH

Ethylene Vinyl Acetate

3vCH

60% Polypropylene/40% Polyester

2vCH

3vCH

60% Polypropylene-Polyethylene copolymer and 40% Polyester

2vCH

3vCH

Atactic Polypropylene

2vCH

3vCH

Absorbance

| Cm⁻¹ | 9000 | 8000 | 7000 | 6000 | 5000 | 4000 |
| Nm | 1111 | 1250 | 1429 | 1667 | 2000 | 2500 |

Spectra used by permission "NIR Spectra of Organic Compounds," © Wiley-VCH, ISBN 3-527-31630-2.

FIGURE A2a.42

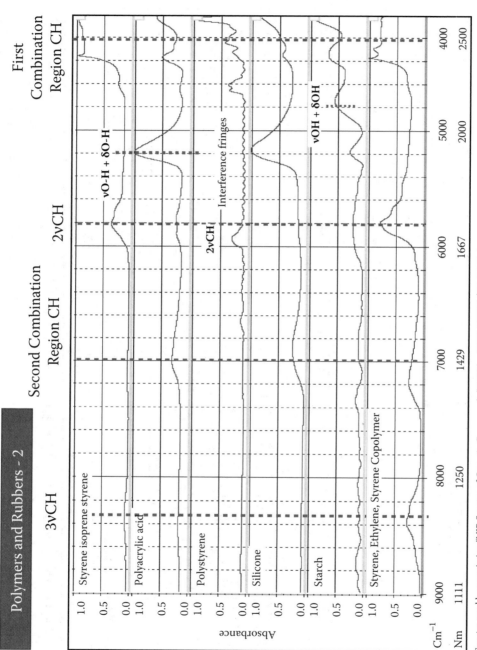

Polymers and Rubbers – 2

Spectra used by permission "NIR Spectra of Organic Compounds," © Wiley-VCH, ISBN 3-527-31630-2.

FIGURE A2a.43

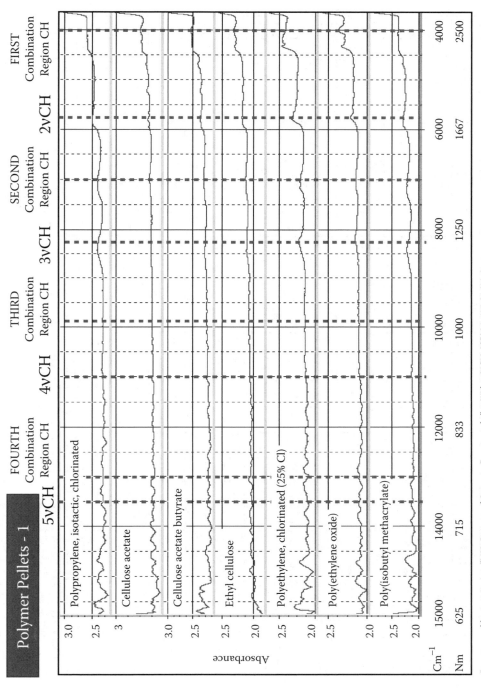

Polymer Pellets – 1

FIRST Combination Region CH
SECOND Combination Region CH
THIRD Combination Region CH
FOURTH Combination Region CH

5vCH 4vCH 3vCH 2vCH

Polypropylene, isotactic, chlorinated
Cellulose acetate
Cellulose acetate butyrate
Ethyl cellulose
Polyethylene, chlorinated (25% Cl)
Poly(ethylene oxide)
Poly(isobutyl methacrylate)

Absorbance

| Cm⁻¹ | 15000 | 14000 | 12000 | 10000 | 8000 | 6000 | 4000 |
| Nm | 625 | 715 | 833 | 1000 | 1250 | 1667 | 2500 |

Spectra used by permission "NIR Spectra of Organic Compounds," © Wiley-VCH, ISBN 3-527-31630-2.

FIGURE A2a.44

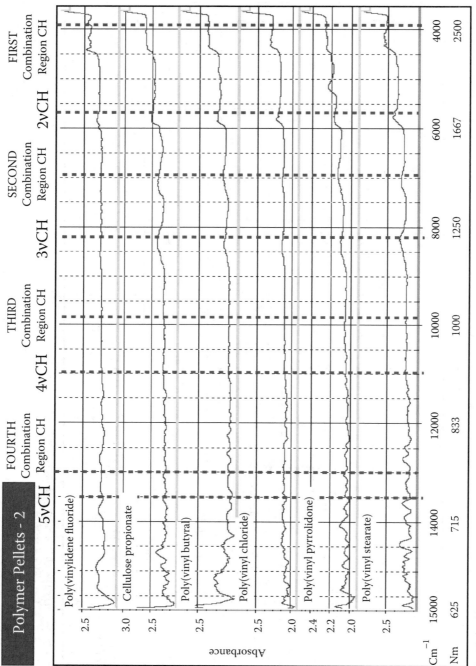

Polymer Pellets - 2

Poly(vinylidene fluoride)
Cellulose propionate
Poly(vinyl butyral)
Poly(vinyl chloride)
Poly(vinyl pyrrolidone)
Poly(vinyl stearate)

Spectra used by permission "NIR Spectra of Organic Compounds," © Wiley-VCH, ISBN 3-527-31630-2.

FIGURE A2a.45

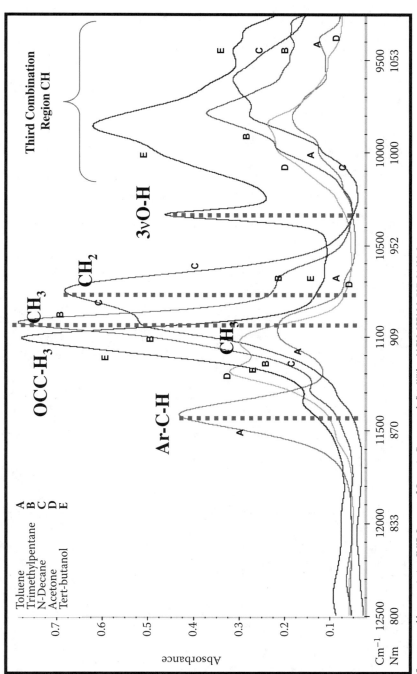

Spectra used by permission "NIR Spectra of Organic Compounds," © Wiley-VCH, ISBN 3-527-31630-2.

FIGURE A2a.46

Appendix 2b: Spectra–Structure Correlations—Labeled Spectra from 7200 cm⁻¹ to 3800 cm⁻¹ (1389 nm to 2632 nm)

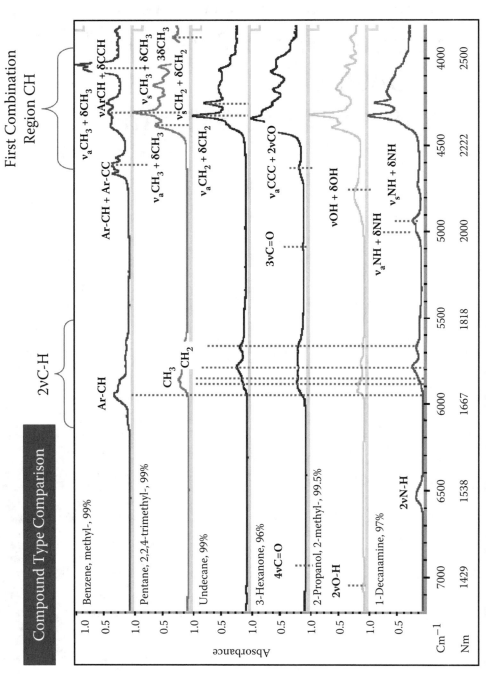

FIGURE A2b.1

Spectra used by permission "NIR Spectra of Organic Compounds," © Wiley-VCH, ISBN 3-527-31630-2.

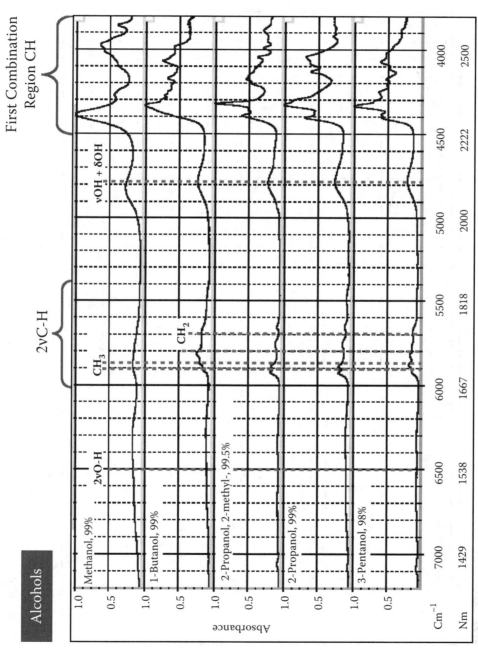

First Combination Region CH

2νC-H

Alcohols

νOH + δOH

CH₃   CH₂

2νO-H

Methanol, 99%

1-Butanol, 99%

2-Propanol, 2-methyl-, 99.5%

2-Propanol, 99%

3-Pentanol, 98%

Absorbance

| Cm⁻¹ | 7000 | 6500 | 6000 | 5500 | 5000 | 4500 | 4000 |
| Nm | 1429 | 1538 | 1667 | 1818 | 2000 | 2222 | 2500 |

Spectra used by permission "NIR Spectra of Organic Compounds," © Wiley-VCH, ISBN 3-527-31630-2.

FIGURE A2b.2

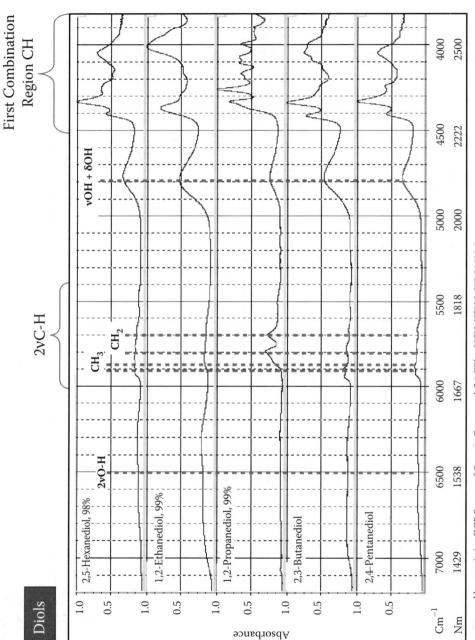

FIGURE A2b.3

Spectra used by permission "NIR Spectra of Organic Compounds," © Wiley-VCH, ISBN 3-527-31630-2.

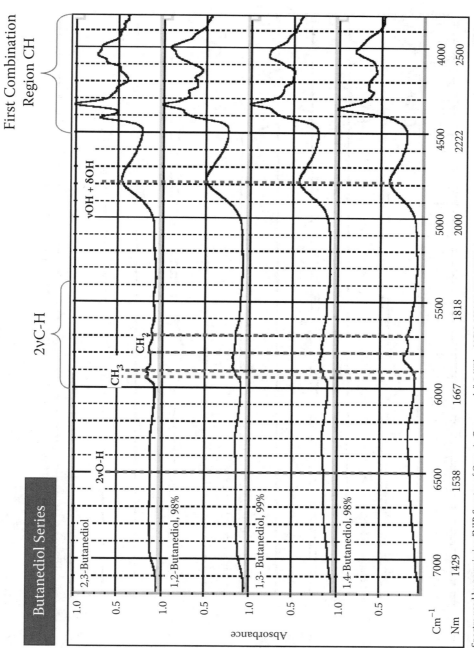

Spectra used by permission "NIR Spectra of Organic Compounds," © Wiley-VCH, ISBN 3-527-31630-2.

FIGURE A2b.4

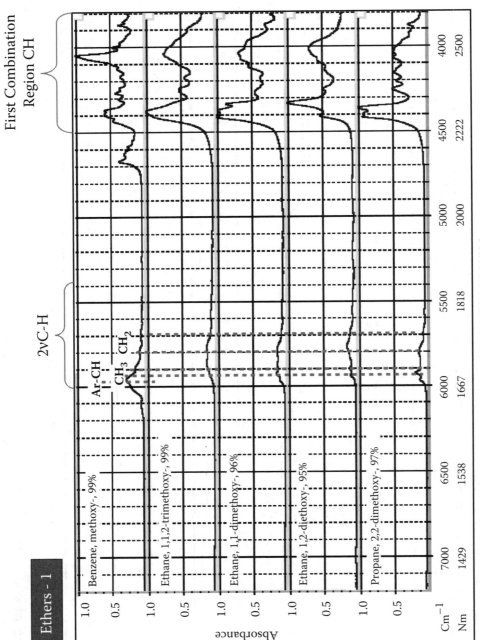

Spectra used by permission "NIR Spectra of Organic Compounds," © Wiley-VCH, ISBN 3-527-31630-2.

FIGURE A2b.5

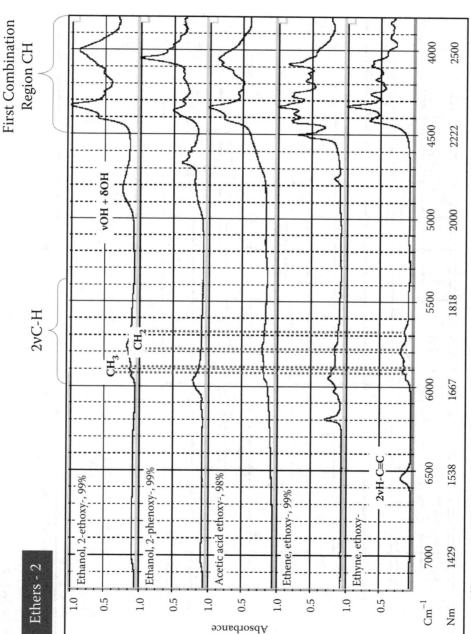

First Combination Region CH

2νC–H

Ethers - 2

νOH + δOH

CH₃   CH₂

2νH–C≡C

Ethanol, 2-ethoxy-, 99%

Ethanol, 2-phenoxy-, 99%

Acetic acid ethoxy-, 98%

Ethene, ethoxy-, 99%

Ethyne, ethoxy-

Absorbance

| Cm⁻¹ | 7000 | 6500 | 6000 | 5500 | 5000 | 4500 | 4000 |
| Nm | 1429 | 1538 | 1667 | 1818 | 2000 | 2222 | 2500 |

Spectra used by permission "NIR Spectra of Organic Compounds," © Wiley-VCH, ISBN 3-527-31630-2.

FIGURE A2b.6

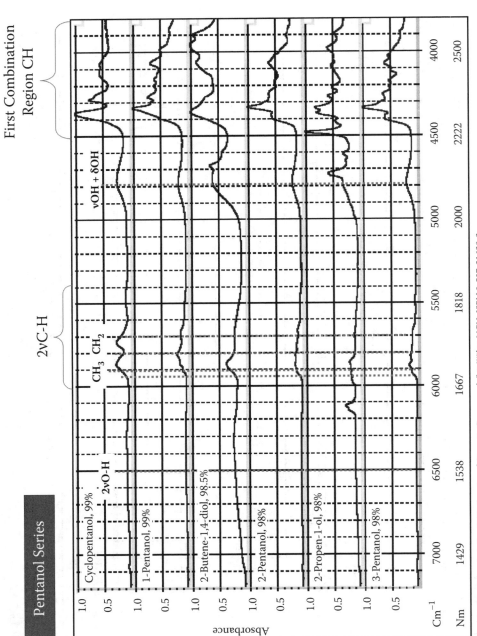

Spectra used by permission "NIR Spectra of Organic Compounds," © Wiley-VCH, ISBN 3-527-31630-2.

FIGURE A2b.7

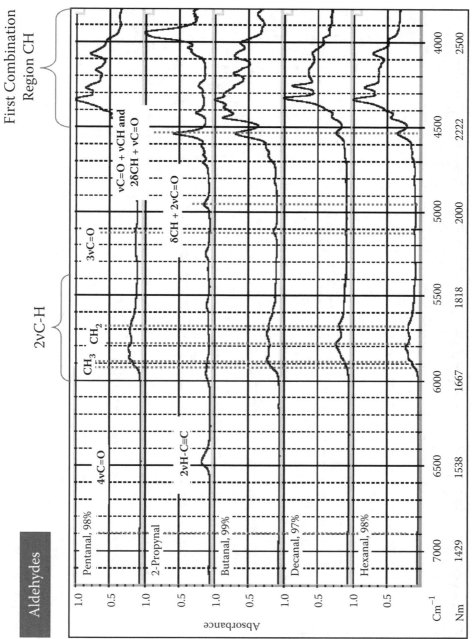

First Combination Region CH

2vC-H

Aldehydes

Pentanal, 98%

4vC=O

2-Propynal

2vH-C≡C

Butanal, 99%

Decanal, 97%

Hexanal, 98%

3vC=O

vC=O + vCH and
2δCH + vC=O

δCH + 2vC=O

CH₃ CH₂

| Cm⁻¹ | 7000 | 6500 | 6000 | 5500 | 5000 | 4500 | 4000 |
| Nm | 1429 | 1538 | 1667 | 1818 | 2000 | 2222 | 2500 |

Absorbance

Spectra used by permission "NIR Spectra of Organic Compounds," © Wiley-VCH, ISBN 3-527-31630-2.

FIGURE A2b.8

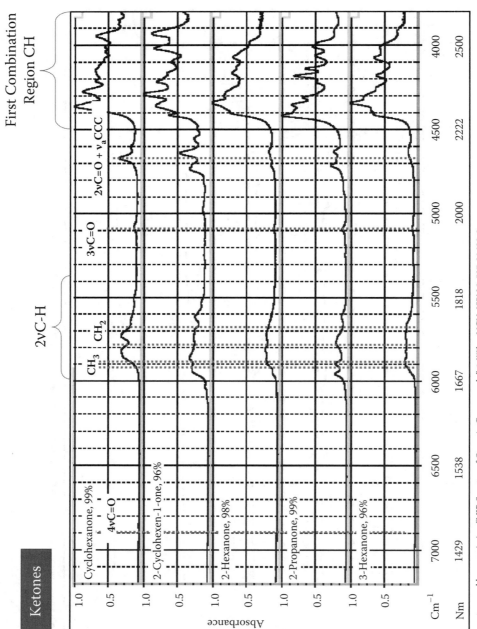

Spectra used by permission "NIR Spectra of Organic Compounds," © Wiley-VCH, ISBN 3-527-31630-2.

FIGURE A2b.9

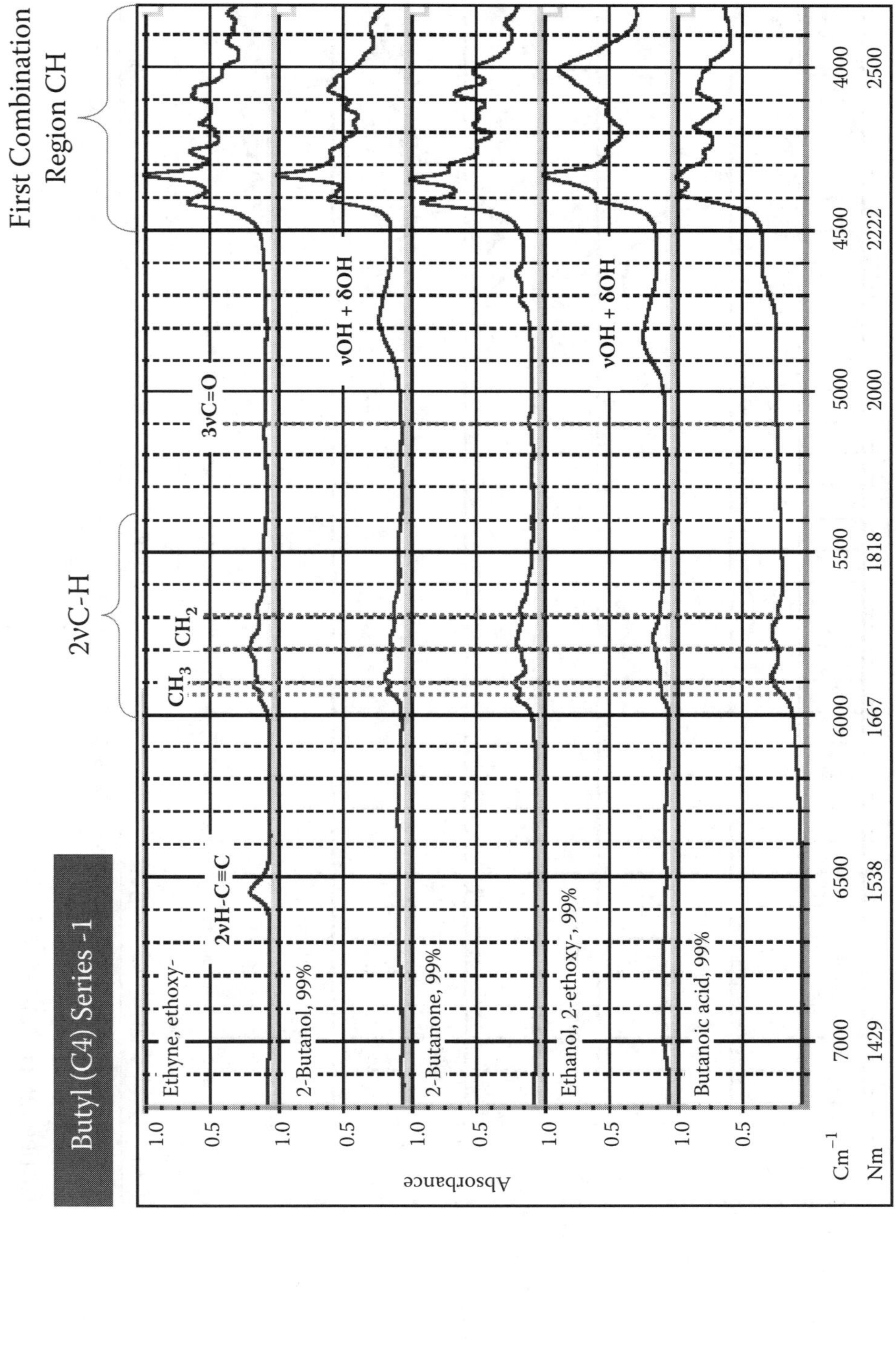
First Combination Region CH
2νC-H
Butyl (C4) Series -1
CH3
CH2
3νC=O
2νH-C≡C
νOH + δOH
νOH + δOH
Ethyne, ethoxy-
2-Butanol, 99%
2-Butanone, 99%
Ethanol, 2-ethoxy-, 99%
Butanoic acid, 99%
Absorbance
1.0
0.5
1.0
0.5
1.0
0.5
1.0
0.5
1.0
0.5
Cm^{-1}
7000
6500
6000
5500
5000
4500
4000
Nm
1429
1538
1667
1818
2000
2222
2500

Spectra used by permission "NIR Spectra of Organic Compounds," © Wiley-VCH, ISBN 3-527-31630-2.

FIGURE A2b.10

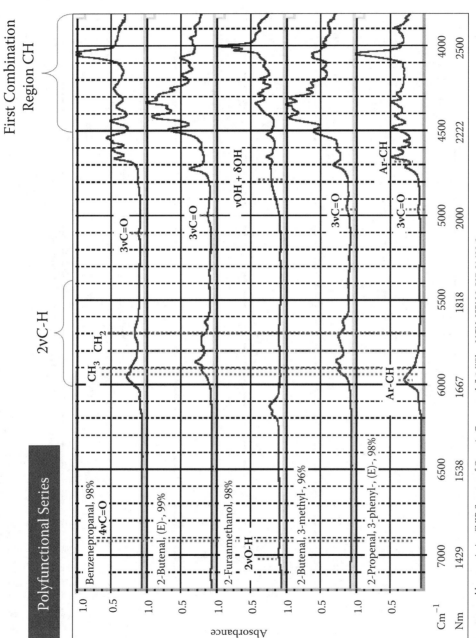

Spectra used by permission "NIR Spectra of Organic Compounds;" © Wiley-VCH, ISBN 3-527-31630-2.

FIGURE A2b.11

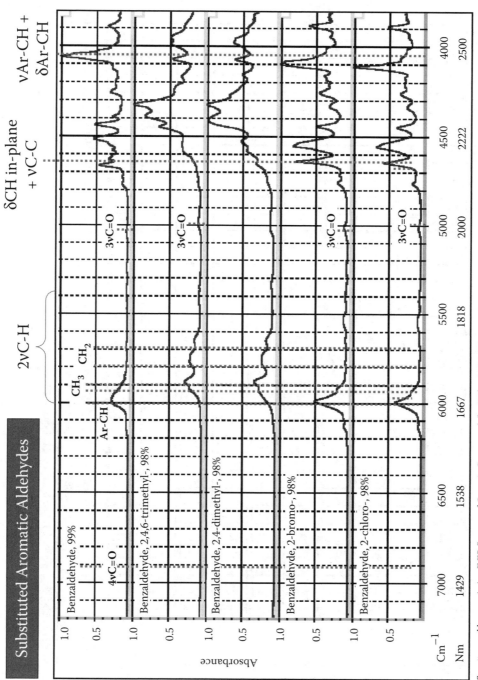

Substituted Aromatic Aldehydes

Spectra used by permission "NIR Spectra of Organic Compounds," © Wiley-VCH, ISBN 3-527-31630-2.

FIGURE A2b.12

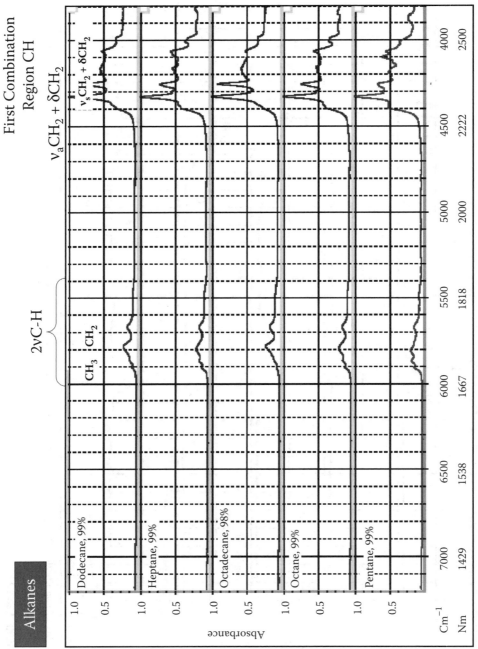

Alkanes

First Combination Region CH

$\nu_a CH_2 + \delta CH_2$

$\nu_s CH_2 + \delta CH_2$

2νC–H

CH_3 CH_2

Dodecane, 99%

Heptane, 99%

Octadecane, 98%

Octane, 99%

Pentane, 99%

Absorbance

| Cm^{-1} | 7000 | 6500 | 6000 | 5500 | 5000 | 4500 | 4000 |
| Nm | 1429 | 1538 | 1667 | 1818 | 2000 | 2222 | 2500 |

Spectra used by permission "NIR Spectra of Organic Compounds," © Wiley-VCH, ISBN 3-527-31630-2.

FIGURE A2b.13

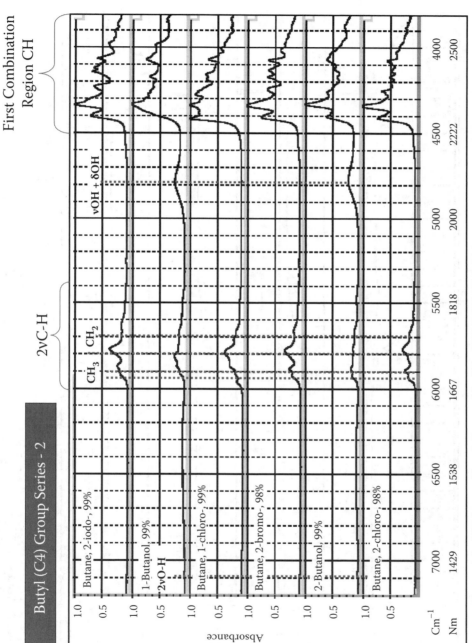

Butyl (C4) Group Series - 2

First Combination Region CH

2vC–H

vOH + δOH

CH₃ CH₂

2vO-H

Butane, 2-iodo-, 99%

1-Butanol, 99%

Butane, 1-chloro-, 99%

Butane, 2-bromo-, 98%

2-Butanol, 99%

Butane, 2-chloro-, 98%

Absorbance

Cm⁻¹	7000	6500	6000	5500	5000	4500	4000
Nm	1429	1538	1667	1818	2000	2222	2500

1.0 0.5

Spectra used by permission "NIR Spectra of Organic Compounds;" © Wiley-VCH, ISBN 3-527-31630-2.

FIGURE A2b.14

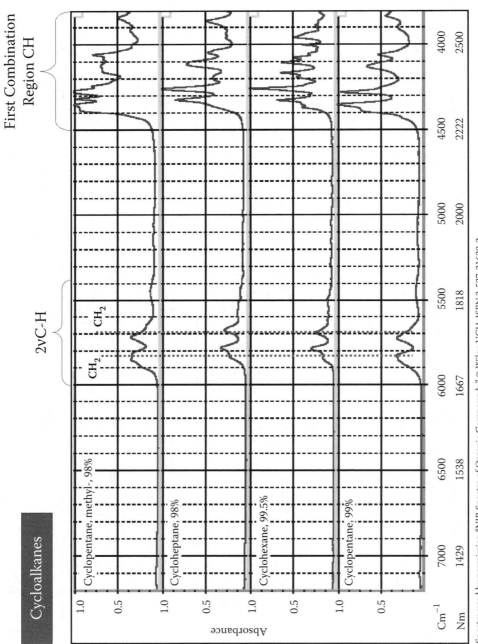

Spectra used by permission "NIR Spectra of Organic Compounds," © Wiley-VCH, ISBN 3-527-31630-2.

FIGURE A2b.15

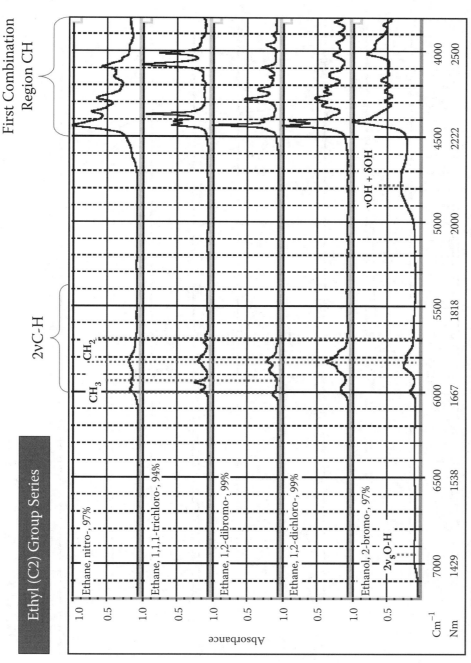

Spectra used by permission "NIR Spectra of Organic Compounds," © Wiley-VCH, ISBN 3-527-31630-2.

FIGURE A2b.16

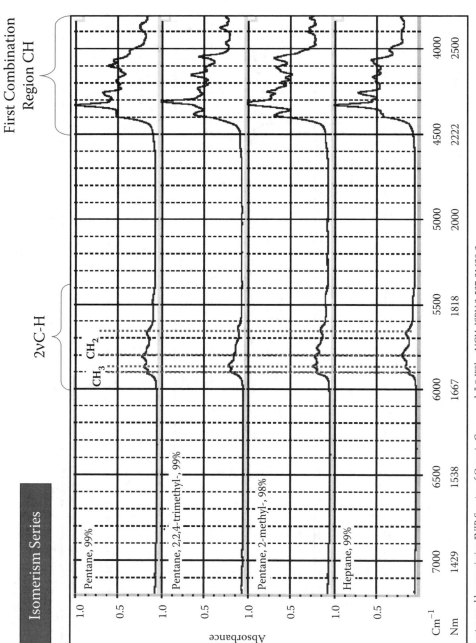

Spectra used by permission "NIR Spectra of Organic Compounds," © Wiley-VCH, ISBN 3-527-31630-2.

FIGURE A2b.17

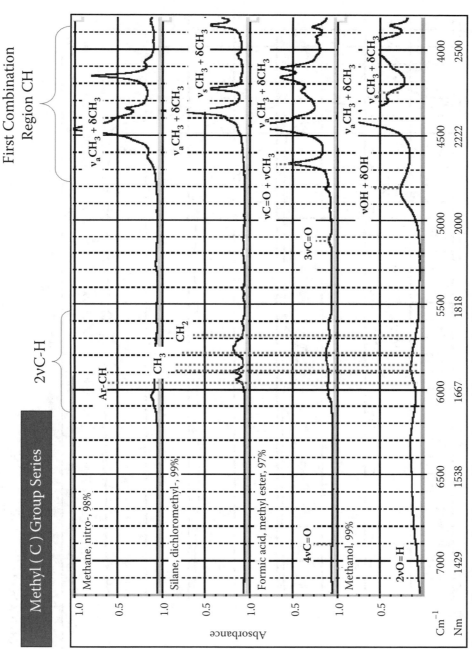

Spectra used by permission "NIR Spectra of Organic Compounds," © Wiley-VCH, ISBN 3-527-31630-2.

FIGURE A2b.18

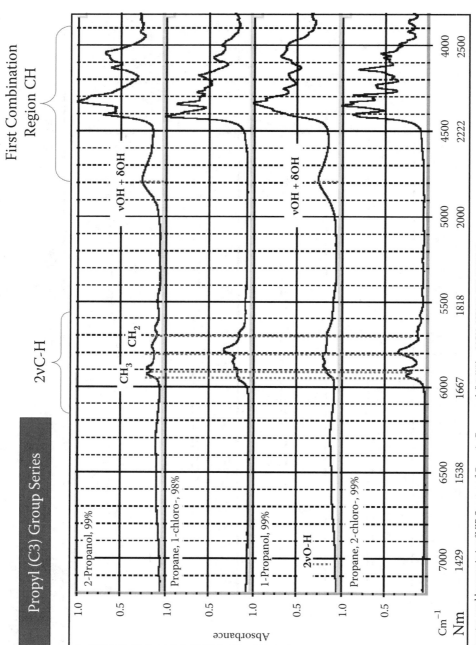

FIGURE A2b.19

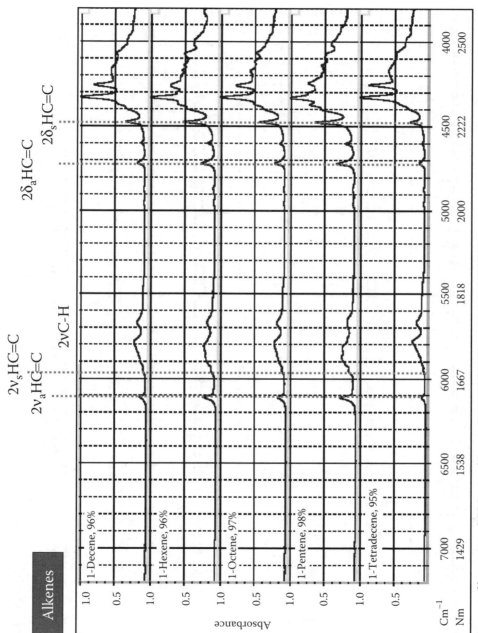

Spectra used by permission "NIR Spectra of Organic Compounds," © Wiley-VCH, ISBN 3-527-31630-2.

FIGURE A2b.20

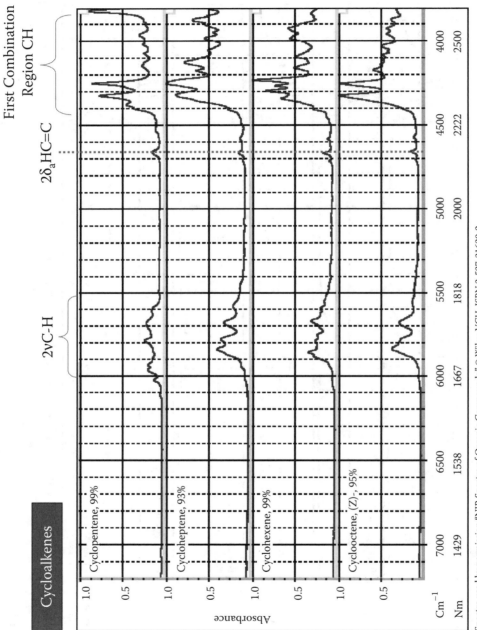

Spectra used by permission "NIR Spectra of Organic Compounds," © Wiley-VCH, ISBN 3-527-31630-2.

FIGURE A2b.21

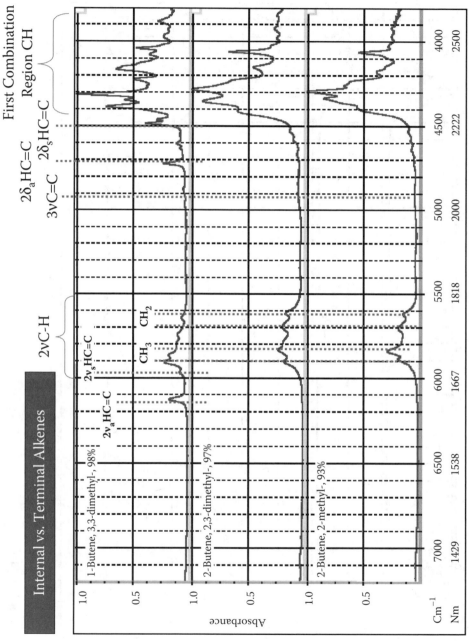

FIGURE A2b.22

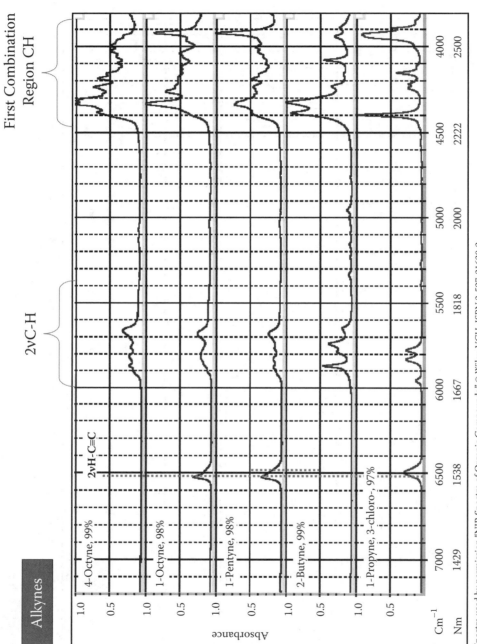

FIGURE A2b.23

Spectra used by permission "NIR Spectra of Organic Compounds," © Wiley-VCH, ISBN 3-527-31630-2.

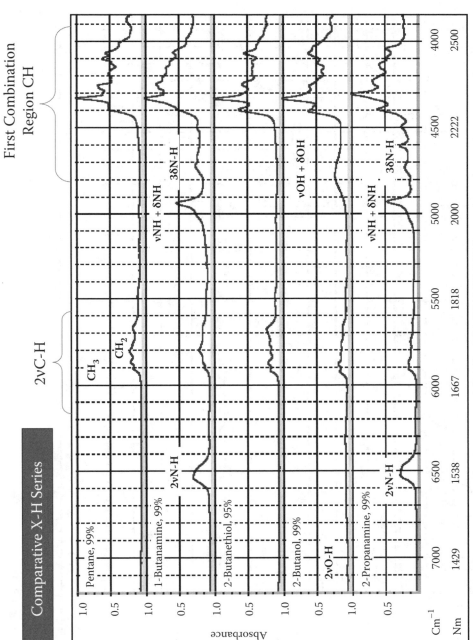

First Combination Region CH

2νC-H

Comparative X-H Series

Pentane, 99%

CH₃

CH₂

1-Butanamine, 99%

νNH + δNH

3δN-H

2νN-H

2-Butanethiol, 95%

2-Butanol, 99%

νOH + δOH

2νO-H

2-Propanamine, 99%

νNH + δNH

3δN-H

2νN-H

| Cm⁻¹ | 7000 | 6500 | 6000 | 5500 | 5000 | 4500 | 4000 |
| Nm | 1429 | 1538 | 1667 | 1818 | 2000 | 2222 | 2500 |

Absorbance

Spectra used by permission "NIR Spectra of Organic Compounds," © Wiley-VCH, ISBN 3-527-31630-2.

FIGURE A2b.24

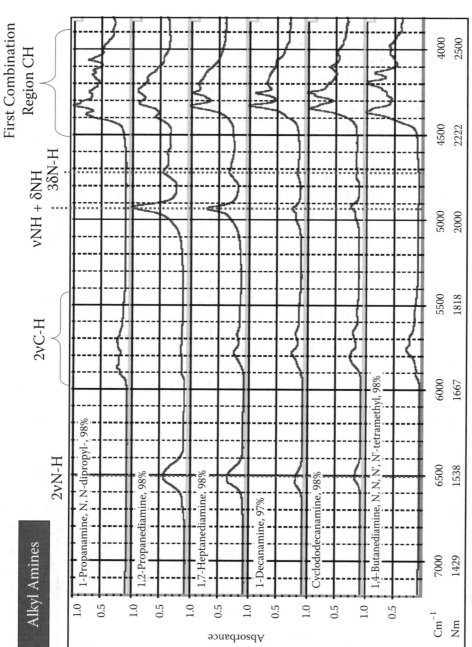

Alkyl Amines

First Combination Region CH

3δN-H

vNH + δNH

2vC-H

2vN-H

1-Propanamine, N, N-dipropyl-, 98%

1,2-Propanediamine, 98%

1,7-Heptanediamine, 98%

1-Decanamine, 97%

Cyclododecanamine, 98%

1,4-Butanediamine, N, N, N', N'-tetramethyl, 98%

Absorbance

| Cm⁻¹ | 7000 | 6500 | 6000 | 5500 | 5000 | 4500 | 4000 |
| Nm | 1429 | 1538 | 1667 | 1818 | 2000 | 2222 | 2500 |

Spectra used by permission "NIR Spectra of Organic Compounds," © Wiley-VCH, ISBN 3-527-31630-2.

FIGURE A2b.25

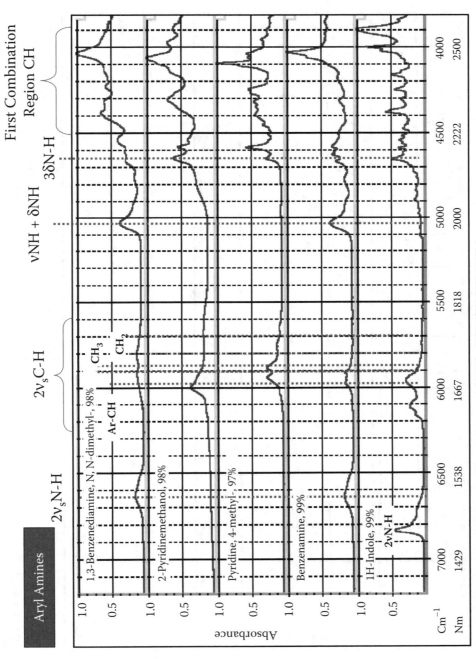

Spectra used by permission "NIR Spectra of Organic Compounds," © Wiley-VCH, ISBN 3-527-31630-2.

FIGURE A2b.26

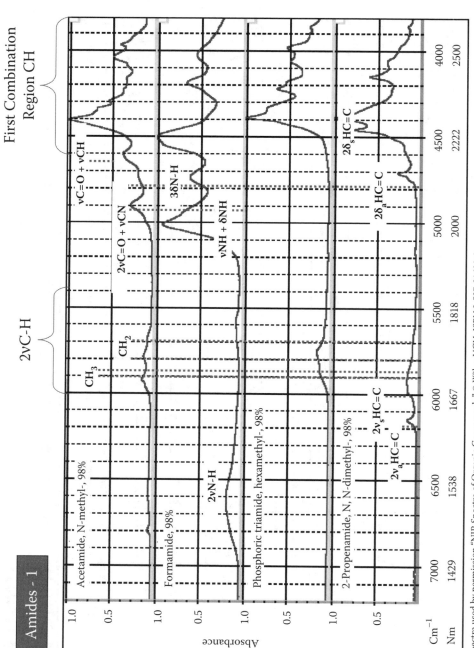

First Combination Region CH

2vC-H

Amides - 1

Acetamide, N-methyl-, 98%

Formamide, 98%

Phosphoric triamide, hexamethyl-, 98%

2-Propenamide, N, N-dimethyl-, 98%

vC=O + vCH
2vC=O + vCN
3δN-H
vNH + δNH
CH₃
CH₂
2vN-H
2δₛHC=C
2δₐHC=C
2νₛHC=C
2νₐHC=C

Absorbance

| Cm⁻¹ | 7000 | 6500 | 6000 | 5500 | 5000 | 4500 | 4000 |
| Nm | 1429 | 1538 | 1667 | 1818 | 2000 | 2222 | 2500 |

Spectra used by permission "NIR Spectra of Organic Compounds," © Wiley-VCH, ISBN 3-527-31630-2.

FIGURE A2b.27

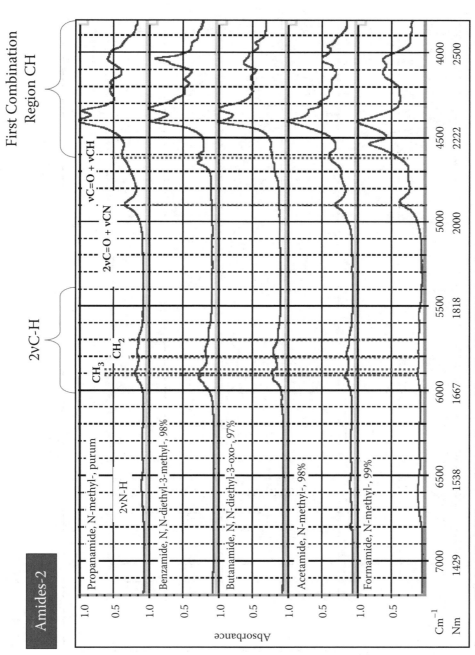

Amides-2

First Combination Region CH

2vC–H

*v*C=O + *v*CH

2*v*C=O + *v*CN

CH₃ CH₂

2*v*N-H

Propanamide, N-methyl-, purum

Benzamide, N, N-diethyl-3-methyl-, 98%

Butanamide, N, N-diethyl-3-oxo-, 97%

Acetamide, N-methyl-, 98%

Formamide, N-methyl-, 99%

Absorbance

Cm⁻¹	7000	6500	6000	5500	5000	4500	4000
Nm	1429	1538	1667	1818	2000	2222	2500

1.0
0.5
1.0
0.5
1.0
0.5
1.0
0.5
1.0
0.5

Spectra used by permission "NIR Spectra of Organic Compounds," © Wiley-VCH, ISBN 3-527-31630-2.

FIGURE A2b.28

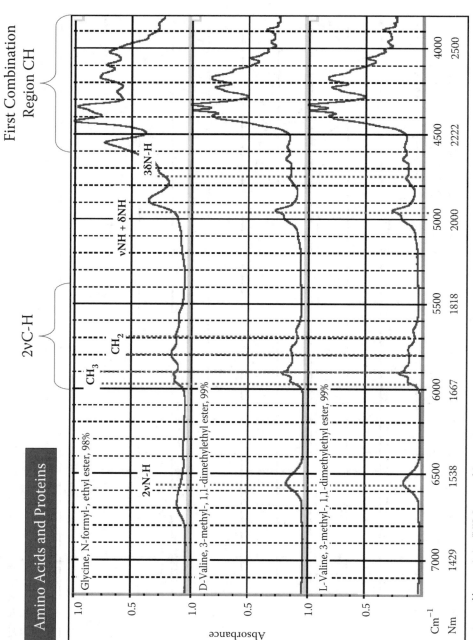

Spectra used by permission "NIR Spectra of Organic Compounds," © Wiley-VCH, ISBN 3-527-31630-2.

FIGURE A2b.29

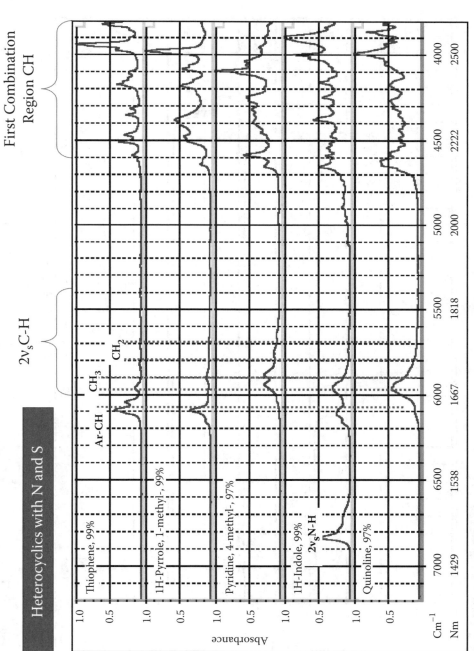

First Combination Region CH

$2\nu_s$C-H

Heterocyclics with N and S

Thiophene, 99%

1H-Pyrrole, 1-methyl-, 99%

Pyridine, 4-methyl-, 97%

1H-Indole, 99%

$2\nu_s$N-H

Quinoline, 97%

Ar-CH

CH_3

CH_2

Absorbance

| Cm^{-1} | 7000 | 6500 | 6000 | 5500 | 5000 | 4500 | 4000 |
| Nm | 1429 | 1538 | 1667 | 1818 | 2000 | 2222 | 2500 |

Spectra used by permission "NIR Spectra of Organic Compounds," © Wiley-VCH, ISBN 3-527-31630-2.

FIGURE A2b.30

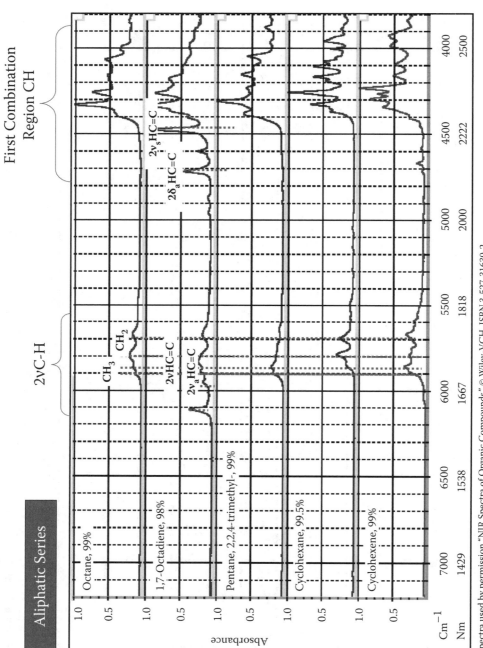

Spectra used by permission "NIR Spectra of Organic Compounds," © Wiley-VCH, ISBN 3-527-31630-2.

FIGURE A2b.31

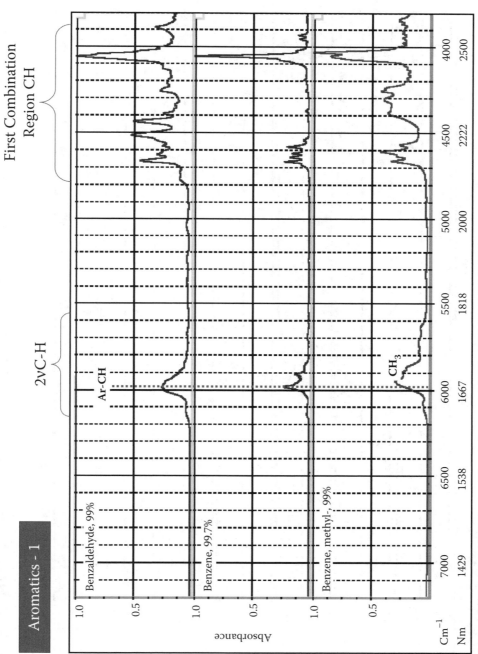

First Combination Region CH

2vC-H

Aromatics - 1

Benzaldehyde, 99%

Ar-CH

Benzene, 99.7%

Benzene, methyl-, 99%

CH₃

Absorbance

| Cm⁻¹ | 7000 | 6500 | 6000 | 5500 | 5000 | 4500 | 4000 |
| Nm | 1429 | 1538 | 1667 | 1818 | 2000 | 2222 | 2500 |

Spectra used by permission "NIR Spectra of Organic Compounds," © Wiley-VCH, ISBN 3-527-31630-2.

FIGURE A2b.32

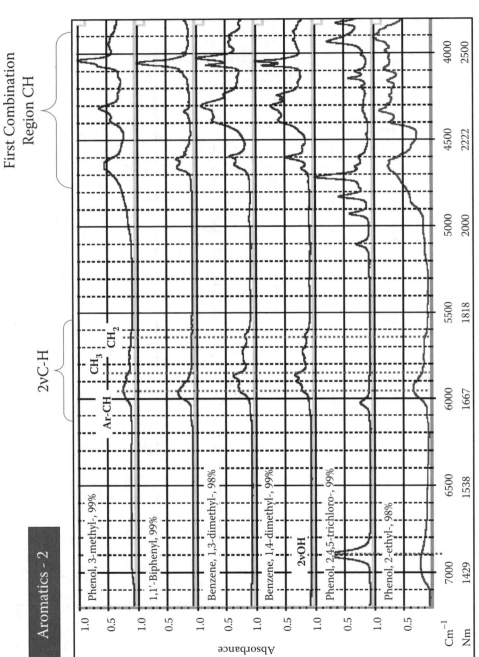

Spectra used by permission "NIR Spectra of Organic Compounds," © Wiley-VCH, ISBN 3-527-31630-2.

FIGURE A2b.33

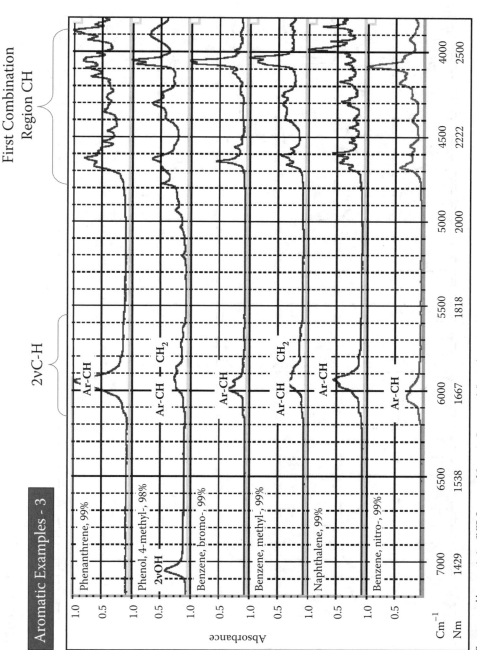

First Combination Region CH

2νC-H

Aromatic Examples - 3

Phenanthrene, 99%

Ar-CH

Phenol, 4-methyl-, 98%

2νOH

Ar-CH + CH$_2$

Benzene, bromo-, 99%

Ar-CH

Benzene, methyl-, 99%

Ar-CH   CH$_2$

Naphthalene, 99%

Ar-CH

Benzene, nitro-, 99%

Ar-CH

Absorbance

| Cm^{-1} | 7000 | 6500 | 6000 | 5500 | 5000 | 4500 | 4000 |
| Nm | 1429 | 1538 | 1667 | 1818 | 2000 | 2222 | 2500 |

Spectra used by permission "NIR Spectra of Organic Compounds," © Wiley-VCH, ISBN 3-527-31630-2.

FIGURE A2b.34

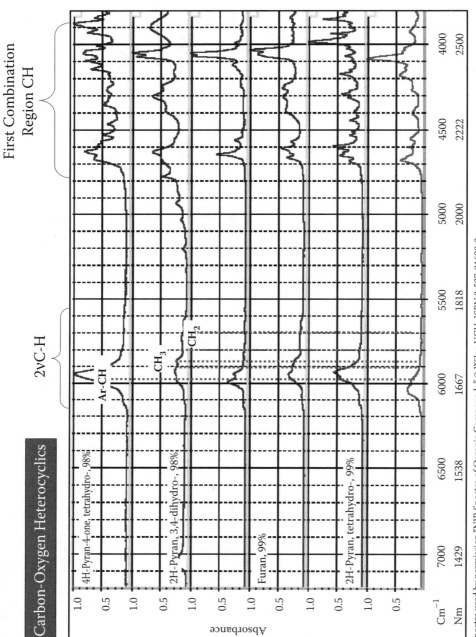

Carbon-Oxygen Heterocyclics

First Combination Region CH

2vC-H

Ar-CH

CH₃

CH₂

4H-Pyran-4-one, tetrahydro-, 98%

2H-Pyran, 3,4-dihydro-, 98%

Furan, 99%

2H-Pyran, tetrahydro-, 99%

Absorbance

| Cm⁻¹ | 7000 | 6500 | 6000 | 5500 | 5000 | 4500 | 4000 |
| Nm | 1429 | 1538 | 1667 | 1818 | 2000 | 2222 | 2500 |

Spectra used by permission "NIR Spectra of Organic Compounds," © Wiley-VCH, ISBN 3-527-31630-2.

FIGURE A2b.35

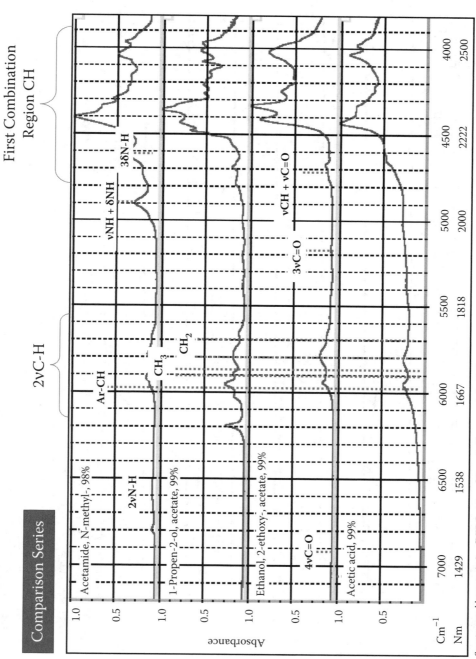

First Combination Region CH

Comparison Series

2vC–H

FIGURE A2b.36

Spectra used by permission "NIR Spectra of Organic Compounds," © Wiley-VCH, ISBN 3-527-31630-2.

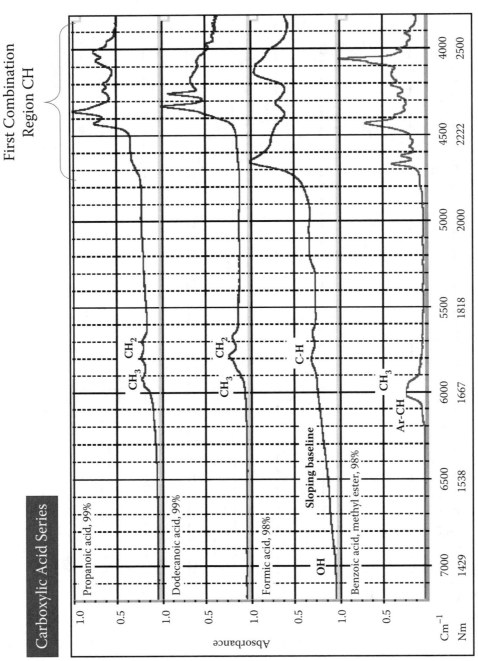

Spectra used by permission "NIR Spectra of Organic Compounds;" © Wiley-VCH, ISBN 3-527-31630-2.

FIGURE A2b.37

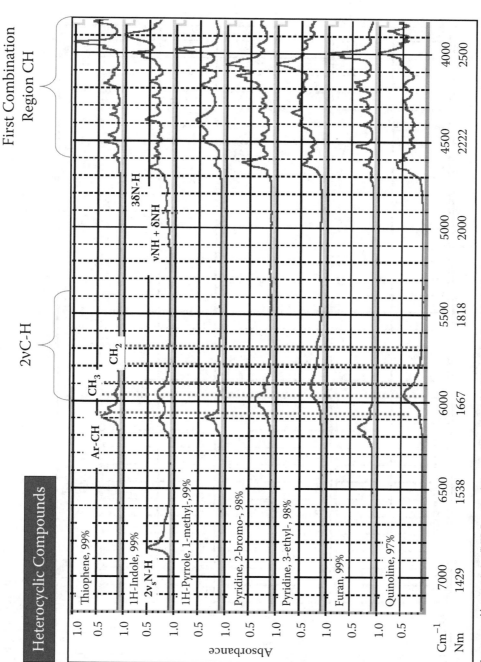

First Combination Region CH

2νC-H

Heterocyclic Compounds

3δN-H

νNH + δNH

CH₂

CH₃

Ar-CH

2νₛN-H

Thiophene, 99%

1H-Indole, 99%

1H-Pyrrole, 1-methyl-, 99%

Pyridine, 2-bromo-, 98%

Pyridine, 3-ethyl-, 98%

Furan, 99%

Quinoline, 97%

Absorbance

| Cm⁻¹ | 7000 | 6500 | 6000 | 5500 | 5000 | 4500 | 4000 |
| Nm | 1429 | 1538 | 1667 | 1818 | 2000 | 2222 | 2500 |

1.0 0.5 1.0 0.5 1.0 0.5 1.0 0.5 1.0 0.5 1.0 0.5 1.0 0.5

Spectra used by permission "NIR Spectra of Organic Compounds," © Wiley-VCH, ISBN 3-527-31630-2.

FIGURE A2b.38

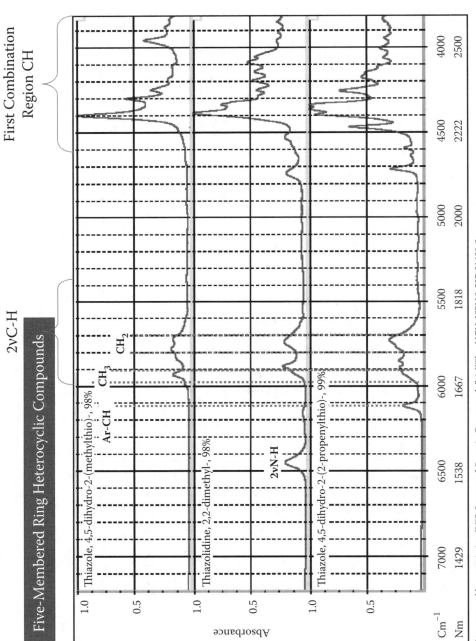

Five-Membered Ring Heterocyclic Compounds

2vC-H

First Combination Region CH

Spectra used by permission "NIR Spectra of Organic Compounds," © Wiley-VCH, ISBN 3-527-31630-2.

FIGURE A2b.39

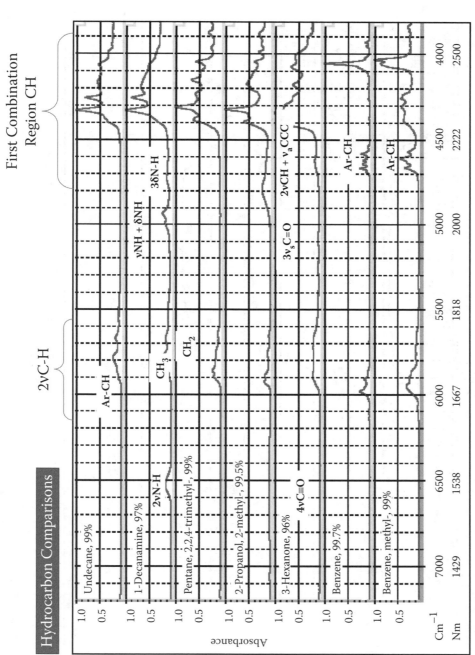

First Combination Region CH

2vC-H

Hydrocarbon Comparisons

Absorbance

| Cm⁻¹ | 7000 | 6500 | 6000 | 5500 | 5000 | 4500 | 4000 |
| Nm | 1429 | 1538 | 1667 | 1818 | 2000 | 2222 | 2500 |

Undecane, 99%
Ar-CH

1-Decanamine, 97%
2vN-H
vNH + δNH
3δN-H

Pentane, 2,2,4-trimethyl-, 99%
CH₃
CH₂

2-Propanol, 2-methyl-, 99.5%

3-Hexanone, 96%
4vC≡O
3v$_s$C=O
2vCH + v$_a$CCC

Benzene, 99.7%
Ar-CH

Benzene, methyl-, 99%
Ar-CH

Spectra used by permission "NIR Spectra of Organic Compounds," © Wiley-VCH, ISBN 3-527-31630-2.

FIGURE A2b.40

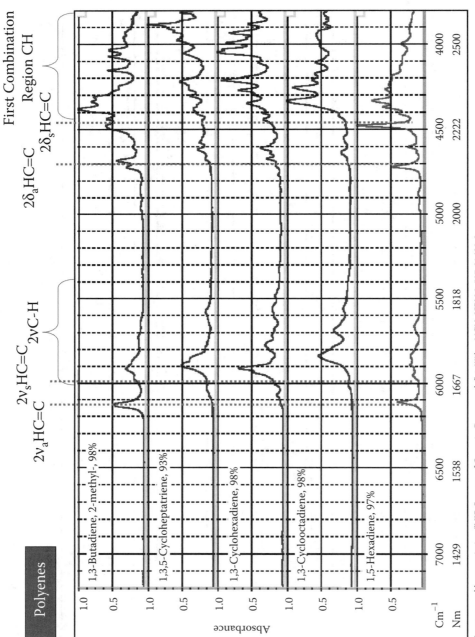

Spectra used by permission "NIR Spectra of Organic Compounds," © Wiley-VCH, ISBN 3-527-31630-2.

FIGURE A2b.41

Appendix 3: Detailed Spectra–Structure Correlations—Overlapping Spectra from 10,500 cm^{-1} to 3800 cm^{-1} (952 nm to 2632 nm)

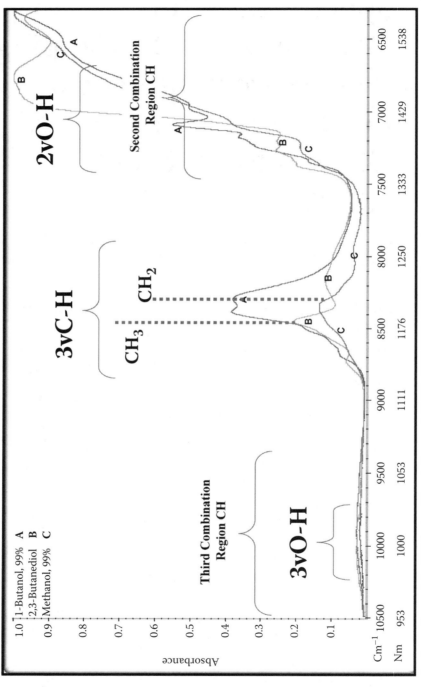

Alcohols

Spectra used by permission "NIR Spectra of Organic Compounds," © Wiley-VCH, ISBN 3-527-31630-2.

FIGURE A3.1

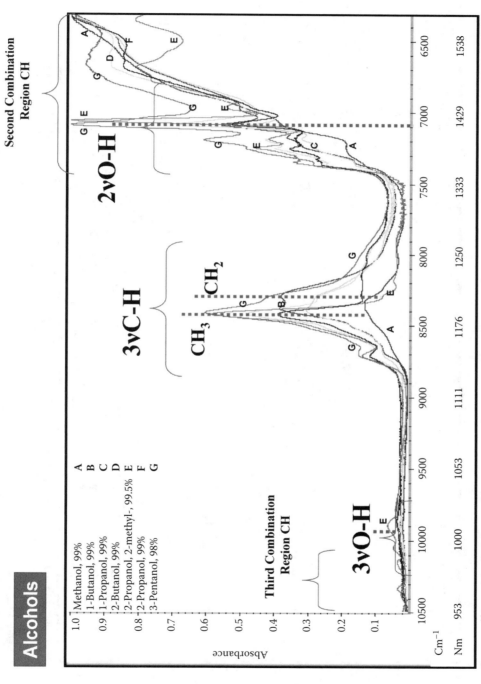

Alcohols

Second Combination Region CH

2νO-H

3νC-H

CH₂

CH₃

Third Combination Region CH

3νO-H

1.0 Methanol, 99%	A
1-Butanol, 99%	B
0.9 1-Propanol, 99%	C
2-Butanol, 99%	D
0.8 2-Propanol, 2-methyl-, 99.5%	E
2-Propanol, 99%	F
3-Pentanol, 98%	G

Absorbance

Cm⁻¹	10500	10000	9500	9000	8500	8000	7500	7000	6500
Nm	953	1000	1053	1111	1176	1250	1333	1429	1538

Spectra used by permission "NIR Spectra of Organic Compounds," © Wiley-VCH, ISBN 3-527-31630-2.

FIGURE A3.2

Alcohols

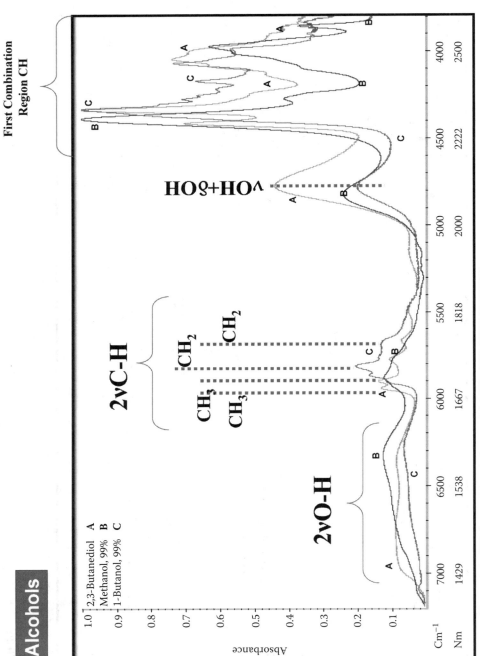

Spectra used by permission "NIR Spectra of Organic Compounds," © Wiley-VCH, ISBN 3-527-31630-2.

FIGURE A3.3

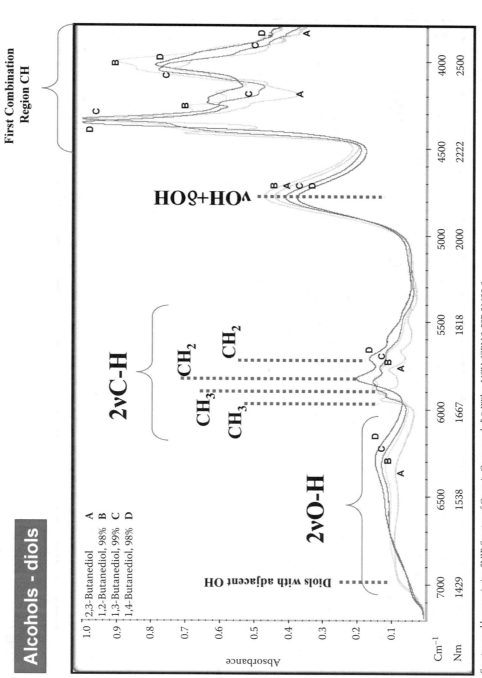

Alcohols - diols

First Combination Region CH

2vC-H

2vO-H

vOH+δOH

CH₃,
CH₃,
CH₂
CH₂

Diols with adjacent OH

2,3-Butanediol   A
1,2-Butanediol, 98%   B
1,3-Butanediol, 99%   C
1,4-Butanediol, 98%   D

Absorbance

| Cm⁻¹ | 7000 | 6500 | 6000 | 5500 | 5000 | 4500 | 4000 |
| Nm | 1429 | 1538 | 1667 | 1818 | 2000 | 2222 | 2500 |

Spectra used by permission "NIR Spectra of Organic Compounds," © Wiley-VCH, ISBN 3-527-31630-2.

FIGURE A3.4

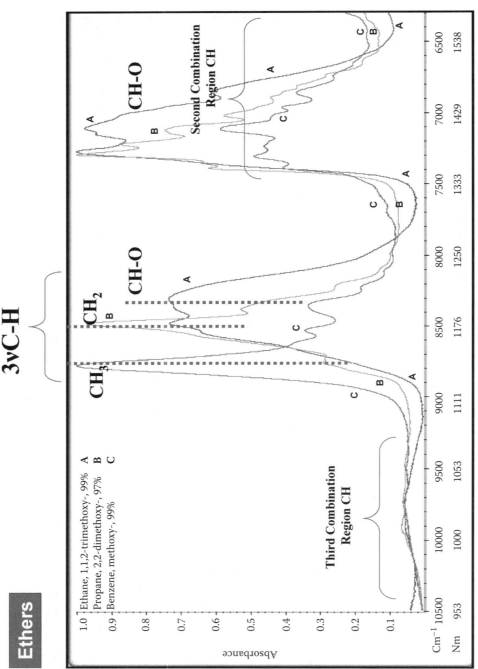

Spectra used by permission "NIR Spectra of Organic Compounds," © Wiley-VCH, ISBN 3-527-31630-2.

FIGURE A3.5

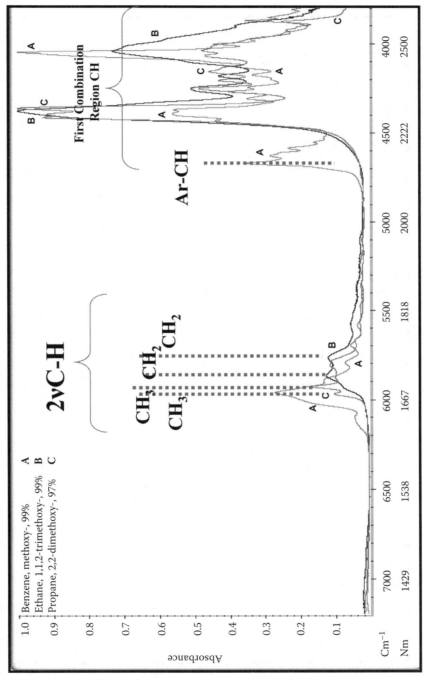

Spectra used by permission "NIR Spectra of Organic Compounds," © Wiley-VCH, ISBN 3-527-31630-2.

FIGURE A3.6

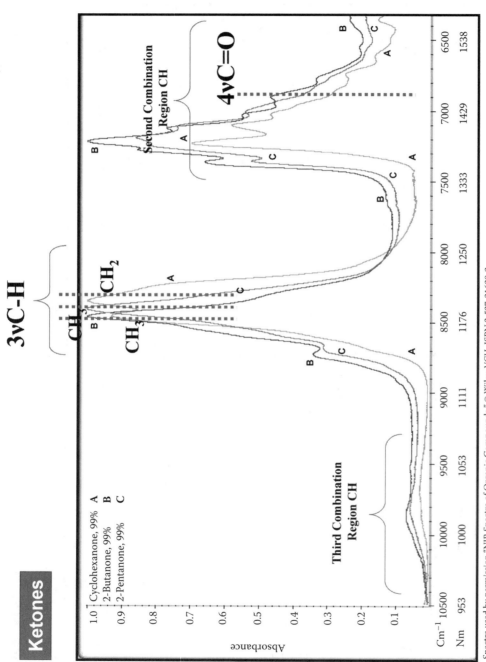

Spectra used by permission "NIR Spectra of Organic Compounds," © Wiley-VCH, ISBN 3-527-31630-2.

FIGURE A3.7

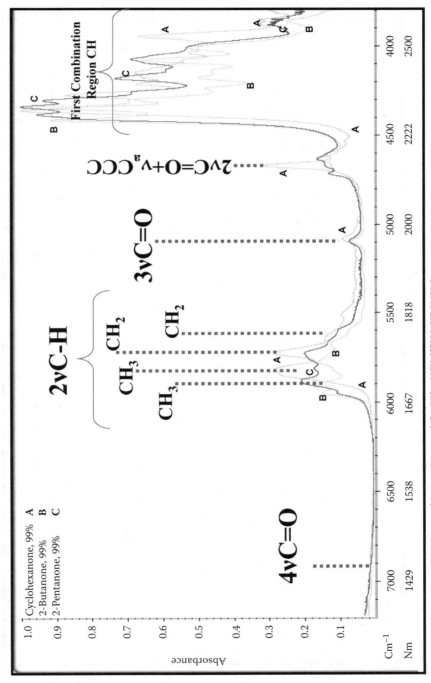

Spectra used by permission "NIR Spectra of Organic Compounds," © Wiley-VCH, ISBN 3-527-31630-2.

FIGURE A3.8

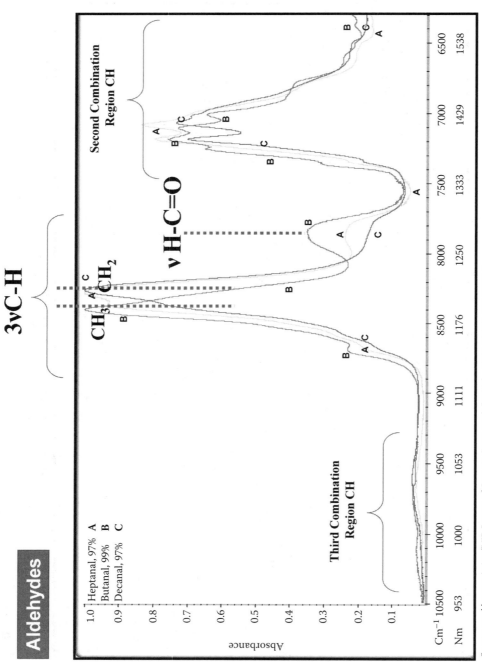

Aldehydes

Spectra used by permission "NIR Spectra of Organic Compounds," © Wiley-VCH, ISBN 3-527-31630-2.

FIGURE A3.9

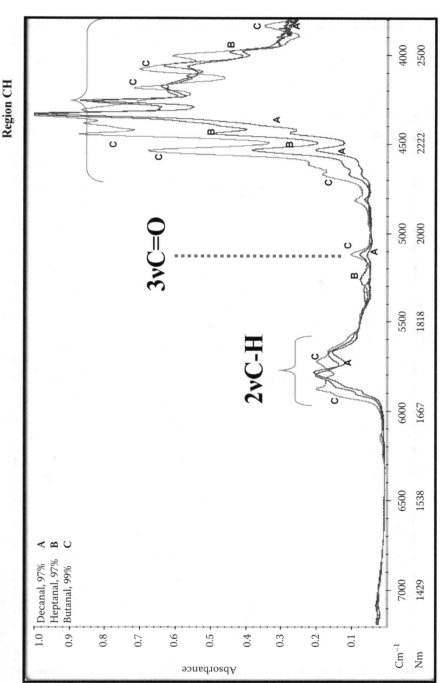

First Combination Region CH

Aldehydes

Decanal, 97% A
Heptanal, 97% B
Butanal, 99% C

$3\nu C{=}O$

$2\nu C\text{-}H$

Absorbance

Cm⁻¹						

Cm⁻¹ 7000 6500 6000 5500 5000 4500 4000
Nm 1429 1538 1667 1818 2000 2222 2500

FIGURE A3.10

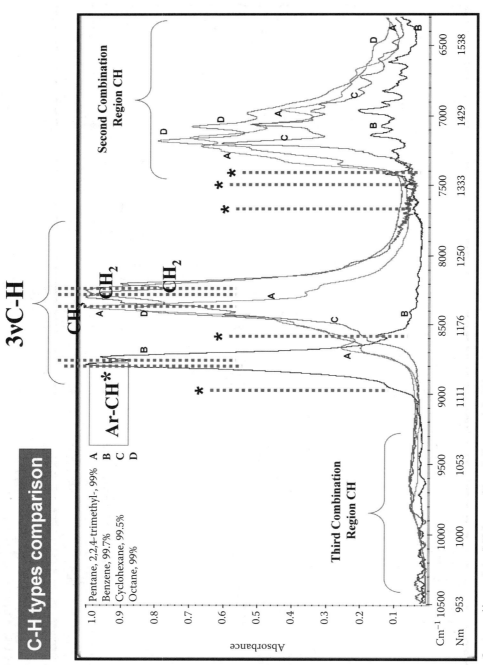

Spectra used by permission "NIR Spectra of Organic Compounds," © Wiley-VCH, ISBN 3-527-31630-2.

FIGURE A3.11

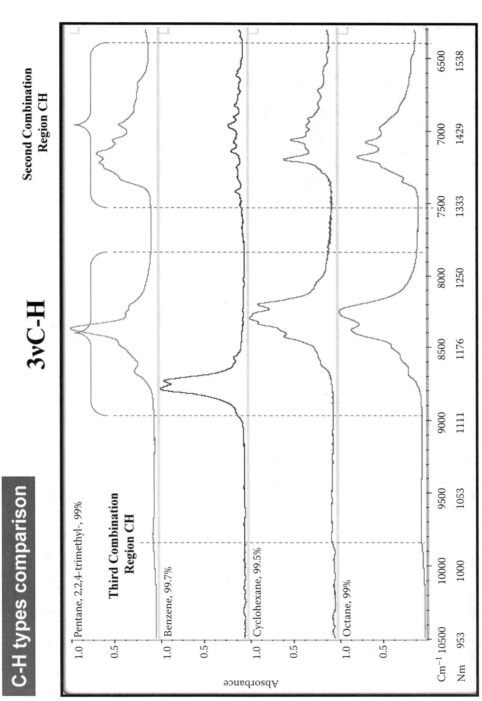

C-H types comparison

3νC-H

Second Combination Region CH

Third Combination Region CH

Pentane, 2,2,4-trimethyl-, 99%

Benzene, 99.7%

Cyclohexane, 99.5%

Octane, 99%

Absorbance

Cm⁻¹	10500	10000	9500	9000	8500	8000	7500	7000	6500
Nm	953	1000	1053	1111	1176	1250	1333	1429	1538

Spectra used by permission "NIR Spectra of Organic Compounds," © Wiley-VCH, ISBN 3-527-31630-2.

FIGURE A3.12

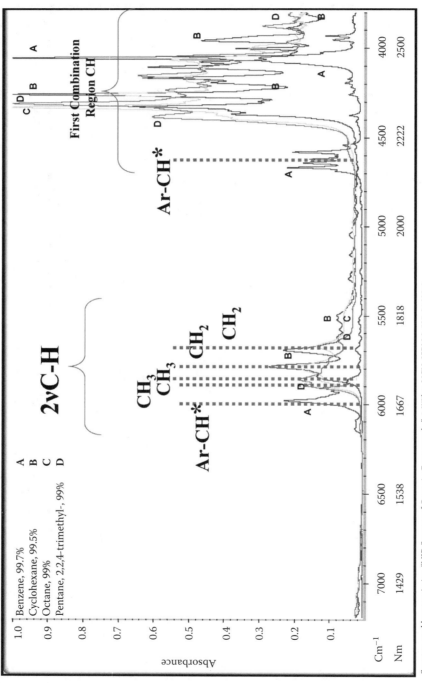

C-H types comparison

A   Benzene, 99.7%
B   Cyclohexane, 99.5%
C   Octane, 99%
D   Pentane, 2,2,4-trimethyl-, 99%

Spectra used by permission "NIR Spectra of Organic Compounds," © Wiley-VCH, ISBN 3-527-31630-2.

FIGURE A3.13

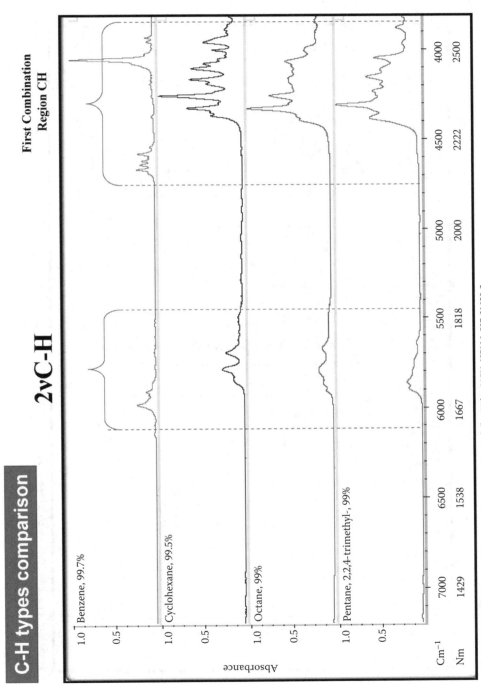

C-H types comparison

2νC-H

First Combination Region CH

Benzene, 99.7%

Cyclohexane, 99.5%

Octane, 99%

Pentane, 2,2,4-trimethyl-, 99%

Absorbance

Cm⁻¹	7000	6500	6000	5500	5000	4500	4000
Nm	1429	1538	1667	1818	2000	2222	2500

Spectra used by permission "NIR Spectra of Organic Compounds," © Wiley-VCH, ISBN 3-527-31630-2.

FIGURE A3.14

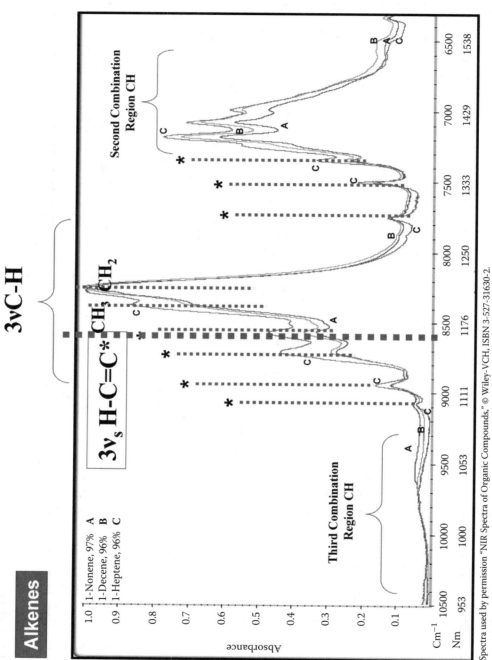

Spectra used by permission "NIR Spectra of Organic Compounds," © Wiley-VCH, ISBN 3-527-31630-2.

FIGURE A3.15

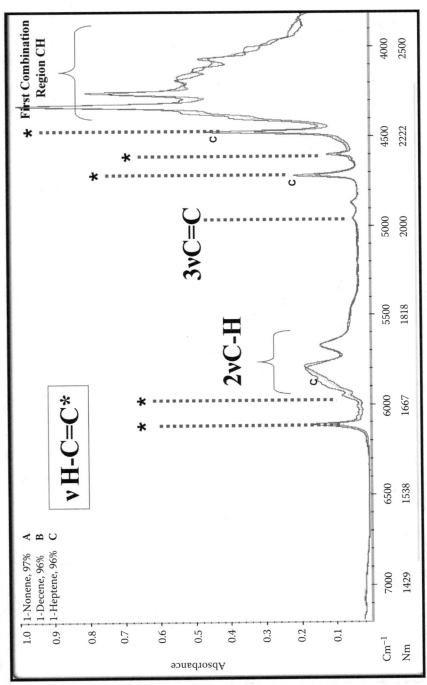

Spectra used by permission "NIR Spectra of Organic Compounds," © Wiley-VCH, ISBN 3-527-31630-2.

FIGURE A3.16

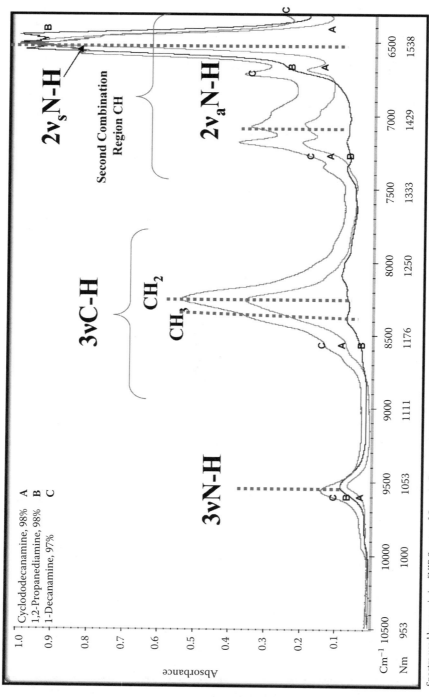

Spectra used by permission "NIR Spectra of Organic Compounds," © Wiley-VCH, ISBN 3-527-31630-2.

FIGURE A3.17

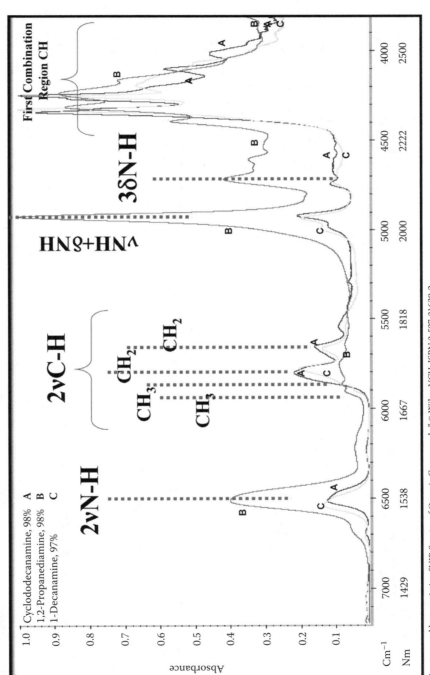

Spectra used by permission "NIR Spectra of Organic Compounds," © Wiley-VCH, ISBN 3-527-31630-2.

FIGURE A3.18

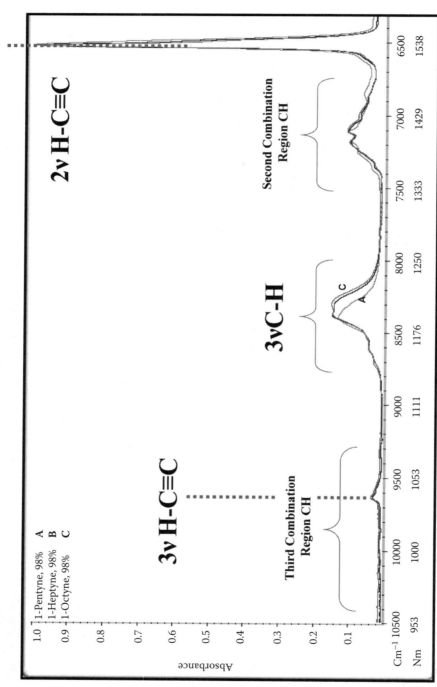

Spectra used by permission "NIR Spectra of Organic Compounds," © Wiley-VCH, ISBN 3-527-31630-2.

FIGURE A3.19

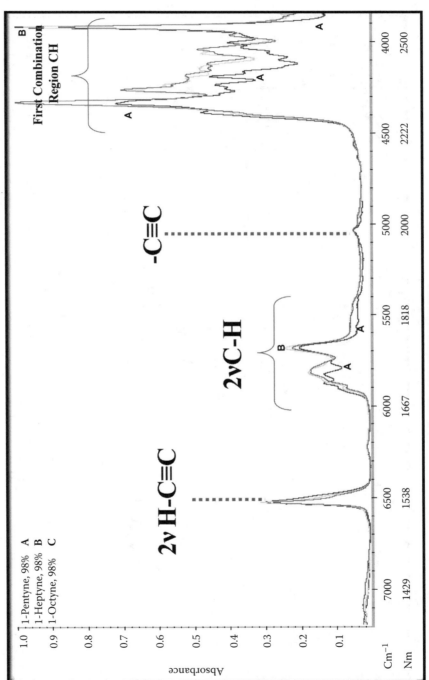

Spectra used by permission "NIR Spectra of Organic Compounds," © Wiley-VCH, ISBN 3-527-31630-2.

FIGURE A3.20

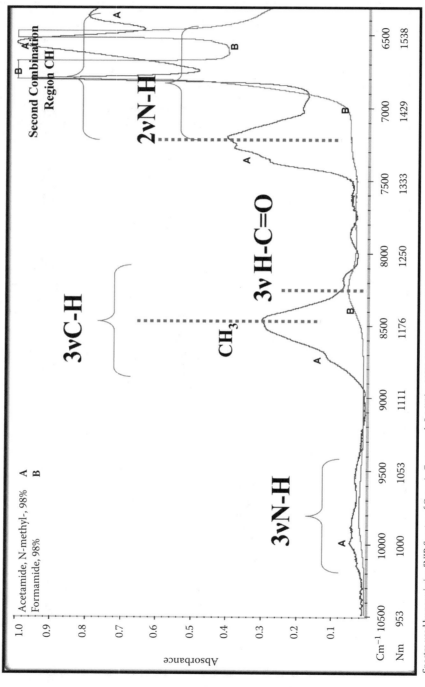

Spectra used by permission "NIR Spectra of Organic Compounds," © Wiley-VCH, ISBN 3-527-31630-2.

FIGURE A3.21

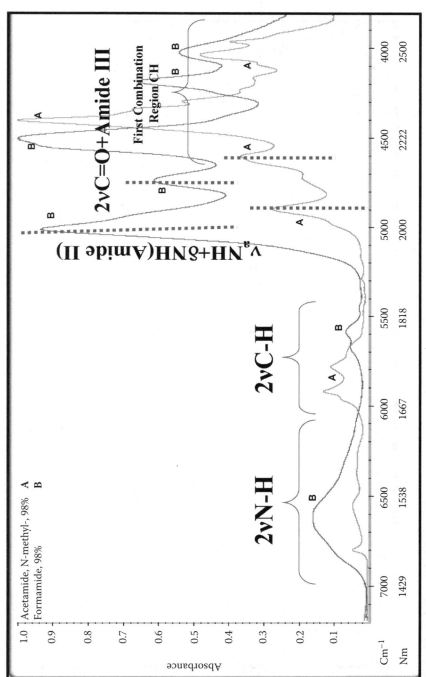

Spectra used by permission "NIR Spectra of Organic Compounds," © Wiley-VCH, ISBN 3-527-31630-2.

FIGURE A3.22

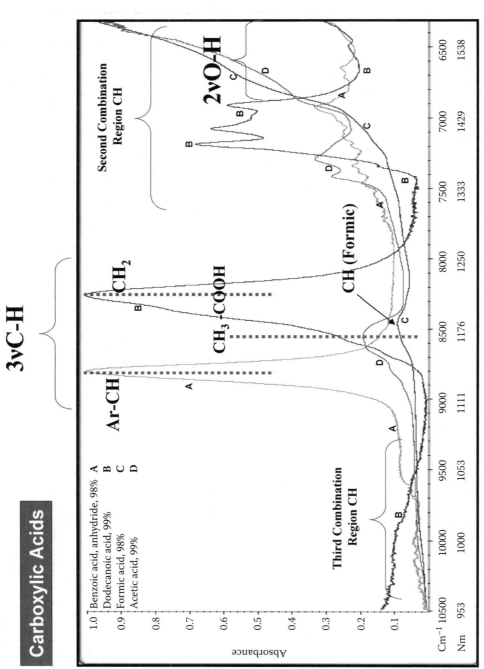

Spectra used by permission "NIR Spectra of Organic Compounds," © Wiley-VCH, ISBN 3-527-31630-2.

FIGURE A3.23

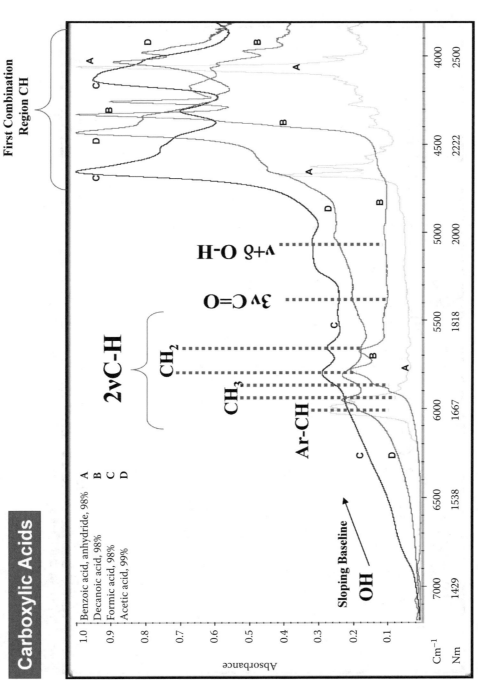

Carboxylic Acids

Benzoic acid, anhydride, 98%	A
Decanoic acid, 98%	B
Formic acid, 98%	C
Acetic acid, 99%	D

First Combination Region CH

2νC-H

CH₂

3ν C=O

ν+β O-H

CH₃

Ar-CH

Sloping Baseline

OH

Spectra used by permission "NIR Spectra of Organic Compounds," © Wiley-VCH, ISBN 3-527-31630-2.

FIGURE A3.24

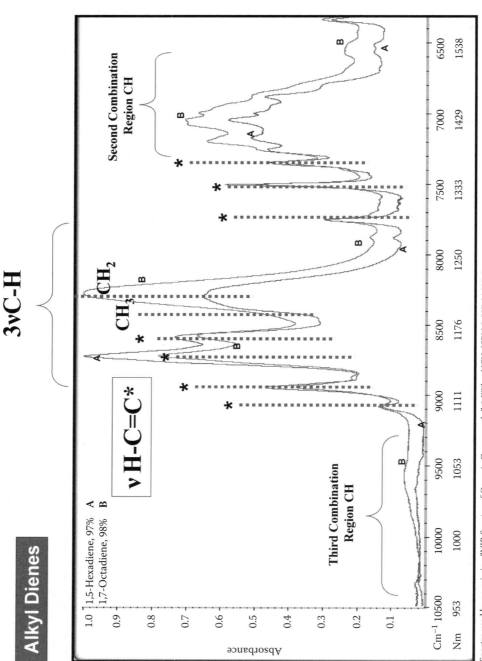

Alkyl Dienes

1,5-Hexadiene, 97% A
1,7-Octadiene, 98% B

3νC-H

ν H-C=C*

CH₂
CH₃

Second Combination Region CH

Third Combination Region CH

Spectra used by permission "NIR Spectra of Organic Compounds," © Wiley-VCH, ISBN 3-527-31630-2.

FIGURE A3.25

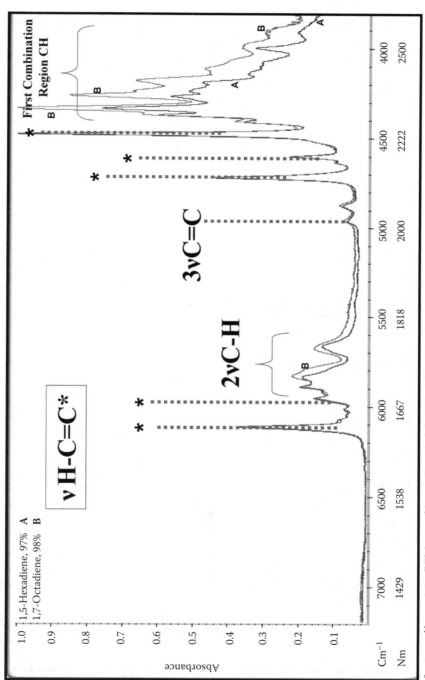

Spectra used by permission "NIR Spectra of Organic Compounds," © Wiley-VCH, ISBN 3-527-31630-2.

FIGURE A3.26

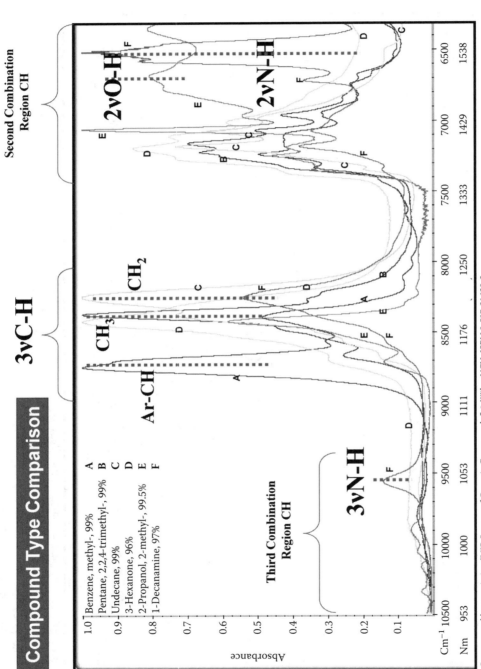

Spectra used by permission "NIR Spectra of Organic Compounds," © Wiley-VCH, ISBN 3-527-31630-2.

FIGURE A3.27

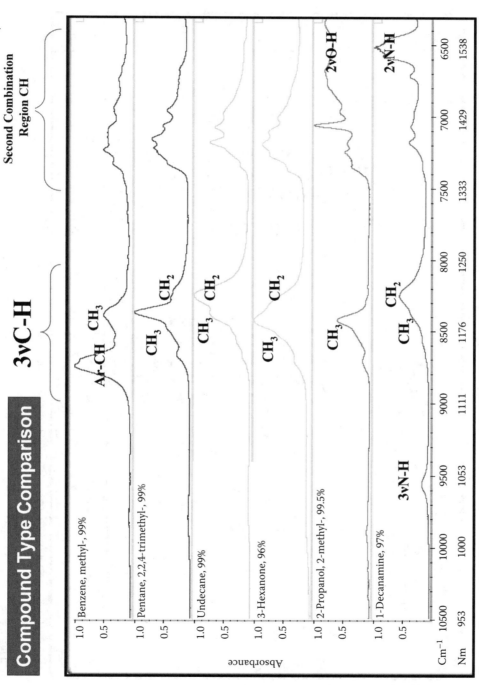

Compound Type Comparison

3νC–H

Second Combination Region CH

Benzene, methyl-, 99%

Ar–CH   CH₃

Pentane, 2,2,4-trimethyl-, 99%

CH₃   CH₂

Undecane, 99%

CH₃   CH₂

3-Hexanone, 96%

CH₃   CH₂

2-Propanol, 2-methyl-, 99.5%

CH₃   CH₂   2νO–H

1-Decanamine, 97%

CH₃   CH₂   2νN–H

3νN–H

Absorbance

Cm⁻¹	10500	10000	9500	9000	8500	8000	7500	7000	6500
Nm	953	1000	1053	1111	1176	1250	1333	1429	1538

Spectra used by permission "NIR Spectra of Organic Compounds," © Wiley-VCH, ISBN 3-527-31630-2.

FIGURE A3.28

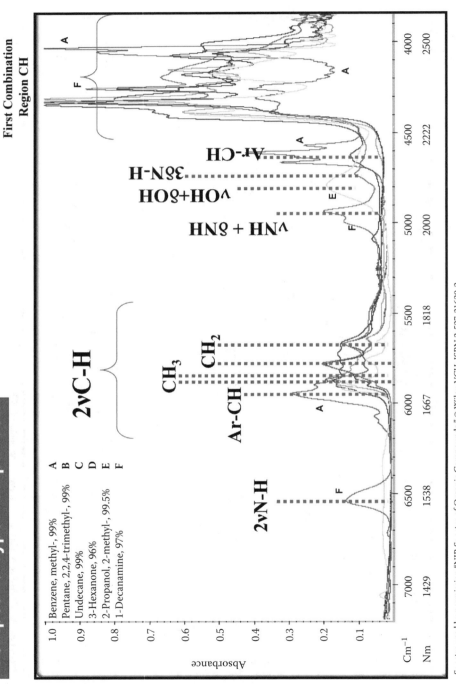

Spectra used by permission "NIR Spectra of Organic Compounds," © Wiley-VCH, ISBN 3-527-31630-2.

FIGURE A3.29

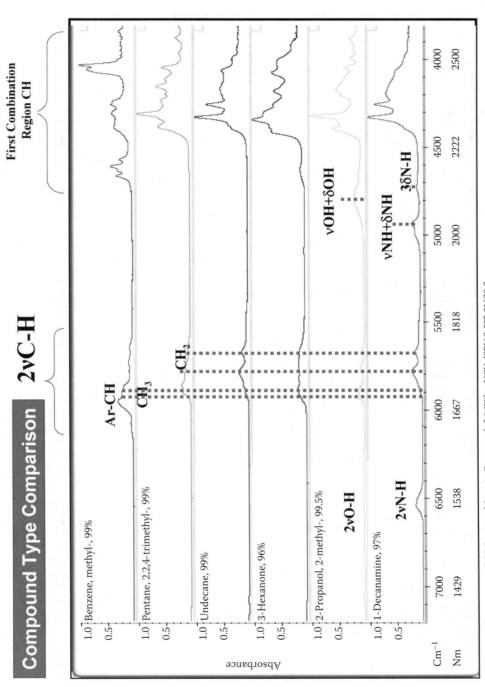

Compound Type Comparison   2νC–H

First Combination Region CH

Ar-CH

CH₃

CH₂

νOH+δOH

νNH+δNH

3δN-H

2νO-H

2νN-H

Benzene, methyl-, 99%						
Pentane, 2,2,4-trimethyl-, 99%						
Undecane, 99%						
3-Hexanone, 96%						
2-Propanol, 2-methyl-, 99.5%						
1-Decanamine, 97%						

Absorbance

| Cm⁻¹ | 7000 | 6500 | 6000 | 5500 | 5000 | 4500 | 4000 |
| Nm | 1429 | 1538 | 1667 | 1818 | 2000 | 2222 | 2500 |

Spectra used by permission "NIR Spectra of Organic Compounds," © Wiley-VCH, ISBN 3-527-31630-2.

FIGURE A3.30

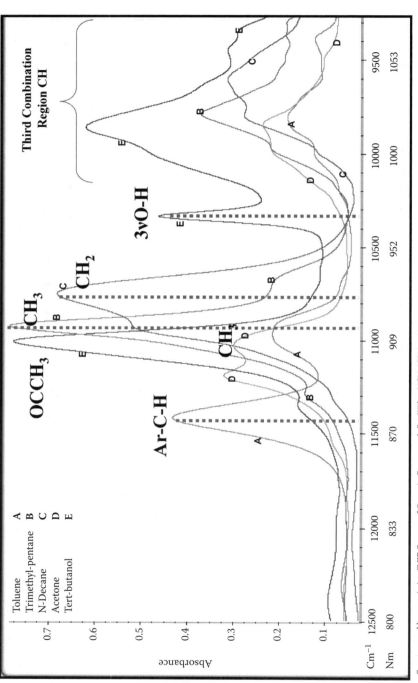

FIGURE A3.31

Spectra used by permission "NIR Spectra of Organic Compounds," © Wiley-VCH, ISBN 3-527-31630-2.

Appendix 4a: Spectra–Structure Correlations for Near Infrared (in Ascending Wavelength and Descending Wavenumber Order)*

* Data for this appendix is referenced within the chapters of this book. Additional limited data is reprinted from Burns, Don and Ciurczak, Emil (eds.), *Handbook of Near Infrared Analysis*, 2nd ed., Marcel Dekker, 2001, pp. 431–433. Used with permission.

Nanometers (nm)	Wavenumbers (cm^{-1})	Functional Group	Spectra–Structure	Material Type
513	19,500	O–H alkyl alcohols O–H with no hydrogen bonding (R–C–OH) in CCl$_4$	O–H (6ν) from nonhydrogen-bonded short-chain alkyl alcohols in CCl$_4$ as (R–C–OH)	Alkyl alcohols
599	16,700	O–H alkyl alcohols O–H with no hydrogen bonding (R–C–OH) in CCl$_4$	O–H (5ν) from nonhydrogen-bonded short-chain alkyl alcohols in CCl$_4$ as (R–C–OH)	Alkyl alcohols
714	13,986	C–H aromatic (ArCH)	C–H (5ν), aromatic C–H	Hydrocarbons, aromatic
741	13,500	O–H alkyl alcohols O–H with no hydrogen bonding (R–C–OH) in CCl$_4$	O–H (4ν) from nonhydrogen-bonded short-chain alkyl alcohols in CCl$_4$ as (R–C–OH)	Alkyl alcohols
747	13,387	C–H methyl C–H (.CH$_3$)	C–H (5ν), methyl C–H	Hydrocarbons, aliphatic
755	13,250	O–H phenols, dilute phenol in CCl$_4$	4νO–H dilute phenol in CCl$_4$	Phenolic O–H
762	13,123	C–H methylene (.CH$_2$)	C–H (5ν), methylene C–H	Hydrocarbons, aliphatic
764	13,400 and 12,788 (doublet)	O–H phenols, dilute phenol in CCl$_4$	4νO–H, *ortho*-iodophenol	Phenolic O–H
767	13,038	O–H from primary alcohols as (–CH$_2$–OH)	O–H (4ν) (–CH–OH), primary alcohols	Primary alcohols
773	12,937	O–H from secondary alcohols as (–CH–OH)	O–H (4ν) (–CH–OH), secondary alcohols	Secondary alcohols
776	12,887	O–H from tertiary alcohols as (–C–OH)	O–H (4ν) (–C–OH), tertiary alcohols	Tertiary alcohols
796	12,563	C–H methyl C–H, associated with linear aliphatic CH$_3$ (CH$_2$)$_N$ CH$_3$	C–H (4νCH$_3$ and δCH$_3$) combination	Hydrocarbons, aliphatic
803	12,453	C–H methyl C–H, associated with aromatic (ArCH$_3$)	C–H (4νCH$_3$ and δCH$_3$) combination	Hydrocarbons, aromatic
813	12,300	C–H methyl C–H, associated with branched aliphatic RC(CH$_3$)$_3$ or RCH(CH$_3$)$_2$	C–H (4νCH$_3$ and δCH$_3$) combination	Hydrocarbons, aliphatic
830	12,048	C–H methylene C–H, associated with linear aliphatic R(CH$_2$)$_N$R	C–H (4νCH$_2$ and δCH$_2$) combination	Hydrocarbons, aliphatic
836	11,962	C–H methylene C–H, associated with branched aliphatic RC(CH$_3$)$_3$ or RCH(CH$_3$)$_2$	C–H (4νCH$_2$ and δCH$_2$) combination	Hydrocarbons, aliphatic
876	11,655	C–H aromatic (ArCH)	C–H (4ν), aromatic C–H	Hydrocarbons, aromatic
908	11,052	C–H methyl (.CH$_3$)	C–H (4ν), .CH$_3$	Hydrocarbons, methyl
915	10,929	C–H methyl C–H (.CH$_3$)	C–H (4ν), methyl C–H	Hydrocarbons, aliphatic
930	10,753	C–H methylene (.CH$_2$)	C–H (4ν), methylene C–H	Hydrocarbons, aliphatic
930	10,800	C–H methylene (.CH$_2$)	C–H (4ν), .CH$_2$	Hydrocarbons, methylene

962	10,400	O–H alkyl alcohols O–H with no hydrogen bonding (R–C–OH) in CCl_4	O–H (3v) from nonhydrogen-bonded short-chain alkyl alcohols in CCl_4 as (R–C–OH)	Alkyl alcohols
990	9911 and 10,288 (doublet)	O–H phenols, dilute phenol in CCl_4	3vO–H, *ortho*-iodophenol	Phenolic O–H
996	10,040	O–H from primary alcohols as (–CH_2–OH)	O–H (3v) (–CH_2–OH), Primary alcohols	Primary alcohols
1000	10,000	O–H phenols, dilute phenol in CCl_4	3vO–H dilute phenol in CCl_4	Phenolic O–H
1003.0	9970	N–H primary aromatic amine (o-NO_2)	N–H (3v) asymmetric, primary aromatic amine in CCl_4 as *ortho*-NO_2 substituent	Aromatic amine
1004	9960	O–H from secondary alcohols as (–CH–OH)	O–H (3v) (–CH–OH), secondary alcohols	Secondary alcohols
1006	9940	O–H from tertiary alcohols as (–C–OH)	O–H (3v) (–C–OH), tertiary alcohols	Tertiary alcohols
1015.0	9852	N–H primary aromatic amine (m-NO_2)	N–H (3v) asymmetric, primary aromatic amine in CCl_4 as *meta*-NO_2 substituent	Aromatic amine
1017.5	9828	N–H primary aromatic amine (o-Cl)	N–H (3v) asymmetric, primary aromatic amine in CCl_4 as *ortho*-Cl grouping	Aromatic amine
1018.5	9818	N–H primary aromatic amine (m-Cl)	N–H (3v) asymmetric, primary aromatic amine in CCl_4 as *meta*-Cl substituent	Aromatic amine
1018.5	9818	N–H primary aromatic amine (o-OCH_3)	N–H (3v) asymmetric, primary aromatic amine in CCl_4 as *ortho*-OCH3 substituent	Aromatic amine
1019.0	9813	N–H primary aromatic amine (p-Cl)	N–H (3v) asymmetric, primary aromatic amine in CCl_4 as *para*-Cl substituent	Aromatic amine
1021	9796	C–H methyl C–H, associated with aromatic (ArCH_3)	C–H (3vCH_3 and δCH_3) combination	Hydrocarbons, aromatic
1021	9794	C–H methyl C–H, associated with branched aliphatic RC(CH_3)$_3$ or RCH(CH_3)$_2$	C–H (3vCH_3 and δCH_3) combination	Hydrocarbons, aliphatic
1021	9795	C–H methyl C–H, associated with linear aliphatic CH_3 (CH_2)$_N$ CH_3	C–H (3vCH_3 and δCH_3) combination	Hydrocarbons, aliphatic

(continued)

Nanometers (nm)	Wavenumbers (cm⁻¹)	Functional Group	Spectra–Structure	Material Type
1021.0	9794	N–H primary aromatic amine (no other substituents)	N–H (3ν) asymmetric, primary aromatic amine in CCl_4, no other substituents	Aromatic amine
1023.0	9775	N–H primary aromatic amine (p-CH₃)	N–H (3ν) asymmetric, primary aromatic amine in CCl_4 as $para$-CH₃ grouping	Aromatic amine
1026.5	9741	N–H primary aromatic amine (p-NH₂)	N–H (3ν) asymmetric, primary aromatic amine in CCl_4 as $para$-NH₂ grouping	Aromatic amine
1029	9720	O–H/C–O polyfunctional alkyl alcohols	2νO–H and 3νC–O combination	Ethers and esters also containing alcohols
1041	9606	C–H methylene C–H, associated with linear aliphatic R(CH₂)ₙR	C–H (3νCH₂ and δCH₂) combination	Hydrocarbons, aliphatic
1042	9600	C–H methylene C–H, associated with branched aliphatic RC(CH₃)₃ or RCH(CH₃)₂	C–H (3νCH₂ and δCH₂) combination	Hydrocarbons, aliphatic
1047	9550	O–H with hydrogen bonding (R–C–OH)	O–H (3ν) from hydrogen-bonded short-chain alkyl compounds	Alkyl alcohols
1065	9386	O–H combination band, alcohols or water	2νO–H and δC–H methyl combination	Alcohols as R–C–O–H
1142	8754	C–H aromatic (ArCH)	C–H (3ν), aromatic C–H	Hydrocarbons, aromatic
1143	8749	C–H (aromatic C–H)	C–H (3ν), Ar.C–H	Hydrocarbons, aromatic
1160	8621	C=O (carbonyl >C=O)	C=O (5ν)	Hydrocarbons, aliphatic
1170	8547	C–H alkene (.HC=CH)	C–H (3ν), .HC=CH	Alkenes, polyenes
1194	8375	C–H methyl C–H, (.CH₃)	C–H (3ν), methyl C–H	Hydrocarbons, aliphatic
1195	8368	C–H methyl (.CH₃)	C–H (3ν), .C–H₃	Hydrocarbons, aliphatic
1211	8258	C–H methylene (.CH₂)	C–H (3ν), methylene C–H	Hydrocarbons, aliphatic
1215	8230	C–H methylene (.CH₂)	C–H (3ν), .C–H₂	Hydrocarbons, aliphatic
1225	8163	C–H secondary or tertiary carbon (CH)	C–H (3ν), .C–H	Hydrocarbons, aliphatic
1360	7353	C–H methyl (.CH₃)	C–H combination, .C–H₃	Hydrocarbons, aliphatic
1370	7300	C–H methyl C–H, associated with aromatic (ArCH₃)	C–H (2νCH₃ and δCH₃) combination	Hydrocarbons, aromatic
1390	7194	C–H methyl C–H, associated with linear aliphatic CH₃ (CH₂)ₙ CH₃	C–H (2νCH₃ and δCH₃) combination	Hydrocarbons, aliphatic
1390	7194	SiOH	SiOH	Silica; optical fibers
1395	7168	C–H methylene (.CH₂)	C–H combination, .C–H₂	Hydrocarbons, aliphatic

7163	1396	C–H methyl C–H, associated with branched aliphatic RC(CH$_3$)$_3$ or RCH(CH$_3$)$_2$	C–H (2νCH$_3$ and δCH$_3$) combination	Hydrocarbons, aliphatic
7100	1408	O–H methanol O–H with hydrogen bonding	O–H (2ν) from nonhydrogen-bonded methanol in CCl$_4$ as (CH$_3$–OH)	O–H from methanol (nonhydrogen bonded)
7092	1410	C–H methylene C–H, associated with linear aliphatic R(CH$_2$)$_N$R	C–H (2νCH$_2$ and δCH$_2$) combination	Hydrocarbons, aliphatic
7092	1410	O–H alcohol (RO–H)	O–H (2ν), O–H	Hydrocarbons, aliphatic
7090	1410	O–H hydroxyl O–H free form (not hydrogen bonded)	O–H, free hydroxyl group in dilute CCl$_4$	O–H not hydrogen bonded
7085	1411	C–H methylene C–H, associated with branched aliphatic RC(CH$_3$)$_3$ or RCH(CH$_3$)$_2$	C–H (2νCH$_2$ and δCH$_2$) combination	Hydrocarbons, aliphatic
7067	1415	C–H methylene (.CH$_2$)	C–H combination, .C–H$_2$	Hydrocarbons, aliphatic
7065	1415	O–H alcoholic O–H, nonhydrogen bonded	O–H (2ν) nonbonded from diols	Alkyl alcohols (diols)
(7060 primarily) 7200–7015	1416	O–H alcohol O–H with hydrogen bonding	O–H (2ν) from butanol	Alkyl alcohols (containing one O–H)
7057	1417	C–H aromatic C–H	C–H combination, Ar.C–H	Hydrocarbons, aromatic
7042	1420	O–H aromatic (ArO–H)	O–H (2ν), O–H	Hydrocarbons, aromatic
7040	1420	O–H, monomeric phenol in CCl$_4$	2νO–H monomeric phenol in CCl$_4$	Phenolic O–H
7140–6940	1420	O–H phenolic O–H	2νO–H from phenols and aryl alcohols Ar-OH	Phenolic O–H
6995	1430	N–H from primary amides (R–C=O–NH$_2$)	N–H (2ν) symmetric from primary amide (R–C=O–NH$_2$)	N–H from primary amides
6982	1432	N–H primary aromatic amine (o-NO$_2$)	N–H (2ν) asymmetric, primary aromatic amine in CCl$_4$ as *ortho*-NO$_2$ substituent	Aromatic amine
6944	1440	C–H methylene (.CH$_2$)	C–H combination, .C–H$_2$	Hydrocarbons, aliphatic
6940	1441	N–H primary aromatic amine (m-NO$_2$)	N–H (2ν) asymmetric, primary aromatic amine in CCl$_4$ as *meta*-NO$_2$ substituent	Aromatic amine

(continued)

Nanometers (nm)	Wavenumbers (cm⁻¹)	Functional Group	Spectra–Structure	Material Type
1441	6940	O–H from sugar as crystalline sucrose	O–H (2v) carbohydrates (C4 hydroxyl within a crystalline matrix)	Crystalline sucrose
1443.0	6930	N–H primary aromatic amine (o-Cl)	N–H (2v) asymmetric, primary aromatic amine in CCl$_4$ as ortho-Cl grouping	Aromatic amine
1445.0	6920	N–H primary aromatic amine (m-Cl)	N–H (2v) asymmetric, primary aromatic amine in CCl$_4$ as meta-Cl substituent	Aromatic amine
1446	6916	C–H aromatic (ArC–H)	C–H combination, ArC–H	Hydrocarbons, aromatic
1446.0	6916	N–H primary aromatic amine (o-OCH$_3$)	N–H (2v) asymmetric, primary aromatic amine in CCl$_4$ as ortho-OCH3 substituent	Aromatic amine
1448.0	6906	N–H primary aromatic amine (p-Cl)	N–H (2v) asymmetric, primary aromatic amine in CCl$_4$ as para-Cl substituent	Aromatic amine
1448.5	6904	N–H primary aromatic amine (no other substituents)	N–H (2v) asymmetric, primary aromatic amine in CCl$_4$, no other substituents	Aromatic amine
1450	6897	C=O (carbonyl >C=O)	C=O (4v)	Ketones and aldehydes
1450	6897	O–H polymeric (.O–H)	O–H (2v), .O–H	Starch/polymeric alcohol
1452	6887	O–H from primary alcohols as (–CH$_2$–OH)	O–H (2v) (–CH$_2$–OH), primary alcohols	Primary alcohols
1452.5	6885	N–H primary aromatic amine (p-CH$_3$)	N–H (2v) asymmetric, primary aromatic amine in CCl$_4$ as para-CH$_3$ grouping	Aromatic amine
1459.5	6852	N–H primary aromatic amine (p-NH$_2$)	N–H (2v) asymmetric, primary aromatic amine in CCl$_4$ as para-NH$_2$ grouping	Aromatic amine
1460	6849	N–H amide .NH or .NH$_2$	N–H (2v), symmetrical	Urea

1460	O–H alcoholic O–H, intramolecularly bonded	6850	O–H (2ν) intramolecularly bonded from diols	Alkyl alcohols (diols)
1461	N–H for secondary amine as (R–NH–R)	6844	N–H (2ν), secondary amine as (R–NH–R), indole, first overtone band intensity comparisons, CCl$_4$ solution	N–H secondary amine
1463	N–H amide .NH or .NH$_2$	6835	N–H (2ν), .CONH$_2$	Amide/protein
1464	O–H from secondary alcohols as (–CH–OH)	6831	O–H (2ν) (–CH–OH), Secondary alcohols	Secondary alcohols
1464	O–H phenols, *ortho*-substituted halogens	6830	2νO–H in *ortho*-substituted halogens on phenols	Phenolic O–H
1465	N–H for secondary amine as (R–NH–R)	6826	N–H (2ν), secondary amine as (R–NH–R), carbazole, first overtone band intensity comparisons, CCl$_4$ solution	N–H secondary amine
1468	O–H from tertiary alcohols as (–C–OH)	6812	O–H (2ν) (–C–OH), tertiary alcohols	Tertiary alcohols
1470	N–H band from urea (NH$_2$–C=O–NH$_2$)	6803	N–H (2ν) asymmetric stretching from urea	N–H from urea
1470	N–H combination band from primary amides (R–C=O–NH$_2$)	6805	N–H (νN–H asymmetric and νN–H symmetric combination) for primary amides	N–H combination band from primary amides
1470	O–H methanol O–H with hydrogen bonding	6800	O–H (2ν) from nonhydrogen-bonded methanol in CCl$_4$ as (CH$_3$–OH)	Methanol
1471	N–H amide with N–R group	6798	N–H (2ν), .CONHR	Amide/protein
1472	N–H primary aromatic amine (*o*–NO$_2$)	6791	N–H (2ν) symmetric, primary aromatic amine in CCl$_4$ as *ortho*-NO$_2$ substituent	Aromatic amine
1480	N–H nonbonded from polyamide 11	6760	N–H (2ν) stretching nonbonded N–H from polyamide 11	Polyamide 11
1481	N–H for secondary amine as (R–NH–R)	6751	N–H (2ν), secondary amine as (R–NH–R), *N*-benzylamine, first overtone band intensity comparisons, CCl$_4$ solution	N–H secondary amine
1483	N–H amide .NH or .NH$_2$	6743	N–H (2ν), .CONH$_2$	Amide/protein

(continued)

Nanometers (nm)	Wavenumbers (cm⁻¹)	Functional Group	Spectra–Structure	Material Type
1485	6736	N–H band from urea (NH_2–C=O–NH_2)	N–H (2v) symmetric stretching from urea	Urea
1486	6729	N–H for secondary amine as (R–NH–R)	N–H (2v), secondary amine as (R–NH–R), diphenylamine, first overtone band intensity comparisons, CCl_4 solution	Secondary amine
1487.0	6725	N–H primary aromatic amine (m-NO_2)	N–H (2v) symmetric, primary aromatic amine in CCl_4 as $meta$-NO_2 substituent	Aromatic amine
1489.5	6713	N–H primary aromatic amine (m-Cl)	N–H (2v) symmetric, primary aromatic amine in CCl_4 as $meta$-Cl substituent	Aromatic amine
1490	6711	N–H amide with N–R group	N–H (2v), .CONHR	Amide/protein
1490	6710	N–H from primary amides (R–C=O–NH_2)	N–H (2v) asymmetric from primary amide (R–C=O–NH_2)	N–H from primary amides
1490	6711	O–H polymeric (.O–H)	O–H (2v), .O–H	Starch/polymeric alcohol
1491.5	6705	N–H primary aromatic amine (o-Cl)	N–H (2v) symmetric, primary aromatic amine in CCl_4 as $ortho$-Cl grouping	Aromatic amine
1491.5	6705	N–H primary aromatic amine (p-Cl)	N–H (2v) symmetric, primary aromatic amine in CCl_4 as $para$-Cl substituent	Aromatic amine
1492	6702	N–H amide .NH or .NH_2	N–H (2v), ArNH_2	Amide/protein
1492.5	6700	N–H primary aromatic amine (o-OCH_3)	N–H (2v) symmetric, primary aromatic amine in CCl_4 as $ortho$-OCH3 substituent	Aromatic amine
1493	6698	N–H from primary and secondary aromatic amine (mixtures of aniline and N-ethylaniline in CCl_4)	N–H (2v) symmetric, primary aromatic amine in CCl_4 as $ortho$-OCH3 substituent	Aromatic amines
1493.0	6698	N–H primary aromatic amine (no other substituents)	N–H (2v) symmetric, primary aromatic amine in CCl_4, no other substituents	Aromatic amine

1496.5	N–H primary aromatic amine (p-CH₃)	N–H (2v) symmetric, primary aromatic amine in CCl₄ as para-CH₃ grouping	Aromatic amine
1500	O–H stretching band, alcohols or water	2vO–H of different aggregates, such as monomers, dimers, and polyols	Alcohols or water O–H
1500	N–H amide .NH or .NH₂	N–H (2v), .NH₂	Amide/protein
1500	N–H combination band from urea (NH₂–C=O–NH₂)	N–H (vN–H asymmetric and vN–H symmetric combination) from urea	N–H from urea
1500	O–H alcohol O–H with hydrogen bonding	O–H (2v) from hydrogen-bonded alkyl alcohols as (R–C–OH)	Alkyl alcohols (containing one O–H)
1502.5	N–H primary aromatic amine (p-NH₂)	N–H (2v) symmetric, primary aromatic amine in CCl₄ as para-NH₂ grouping	Aromatic amine
1510	N–H amide .NH or .NH₂	N–H (2v), .CONH₂	Amide/protein
1515	N–H bonded from polyamide 11	N–H (2v) stretching bonded NH, disordered phase from polyamide 11	Polyamide 11
1520	N–H amide .NH or .NH₂	N–H (2v), .CONH₂	Amide/protein
1520	N–H for secondary amine as (R–NH–R)	N–H (2v), secondary amine as (R–NH–R), dimethylamine (vapor), first overtone band intensity comparisons, CCl₄ solution	Secondary amine
1530	C–H methyne C–H as (R–C–C≡C–H)	C–H, methyne (1-hexyne) as (R–C–C≡C–H)	Acetylene or methyne C–H
1530	N–H amide .NH or .NH₂	N–H (2v), RNH₂	Amide/protein
1530	N–H for secondary amine as (R–NH–R)	N–H (2v), secondary amine as (R–NH–R), morpholine, first overtone band intensity comparisons, CCl₄ solution	N–H secondary amine
1534	N–H ammonia in water	N–H (2v) for NH₃ (ammonia) in water	Ammonia in water
1538	N–H bonded from polyamide 11	N–H (2v) stretching bonded NH, ordered phase from polyamide 11	Polyamide 11
1540	O–H polymeric (.O–H)	O–H (2v), .O–H	Starch/polymeric alcohol

(continued)

Nanometers (nm)	Wavenumbers (cm⁻¹)	Functional Group	Spectra–Structure	Material Type
1545	6471	N–H for secondary amine as (R–NH–R)	N–H (2ν), secondary amine as (R–NH–R), diethylamine, first overtone band intensity comparisons, CCl_4 solution	N–H secondary amine
1550	6450	O–H-bound water and the first overtone of an NH-stretching mode of the amide groups of ovalbumin	O–H-bound water with several hydrogen bonds and the first overtone of an NH-stretching mode of the amide groups of ovalbumin	O–H-bound water of the amide groups of ovalbumin
1550	6540–6250	N–H from secondary amide in proteins	N–H (2ν) stretching from secondary amide in proteins	N–H from protein
1570	6369	N–H amide .NH	N–H (2ν), .CONHR	Amide/protein
1570	6368	N–H bonded combination from polyamide 11	N–H [bonded NH stretching and 2 × amide II deformation (N–H in-plane bending) combination] from polyamide 11	Polyamide 11
1580	6330	O–H combination band, alcohols or water	2νO–H, O–H broad band, hydrogen bonded	Alcohols as R–C–O–H
1583	6319	O–H-stretching band, alkyl alcohols or water	νO–H and δO–H combination reported in the literature from ethylene-vinyl alcohol copolymer spectra, probably better assigned as 2νO–H	Alcohols or water O–H
1598	6256	C=O/N–H combination from polyamide 11	C=O/N–H [amide I (2νₛC=O stretching) and 3 × amide II (N–H in-plane bending) combination] from polyamide 11	Polyamide 11
1613	6200	C–H vinyl C–H attached to >C=O group, such as (CH_2=CH–C=O –)	C–H, vinyl on O (1-ethenyloxybutane) as (CH_2=CH–C=O –)	Vinyl as 1-ethenyl oxybutane (1-ethenyl-methyl-ethyl-ketone)
1618	6180	N–H from polyamide 11	N–H [4 × amide II (N–H in-plane bending)] from polyamide 11	Polyamide 11
1620	6173	C–H alkene, .=CH_2	C–H (2ν), .=CH_2	Alkenes

1621	6169	C–H acrylate C–H as (CH₂=CHCOO⁻)	C–H, combination of two stretching modes	Acrylate
1621	6170	C–H vinyl C–H attached to N group, such as (CH₂=CH–N–C=O–C)	C–H, vinyl on N (1-ethenyl 2-pyrrolidinone) as (CH₂=CH–N–C=O–C)	Vinyl on pyrrolidinone
1630	6130/6140	C–H vinyl and vinylidene C–H as (CH₂=C(CH₃)–CH=CH₂)	C–H, vinyl and vinylidene (2-methyl-1,3-butadiene), isoprene as (CH₂=C(CH₃)–CH=CH₂)	Vinyl and vinylidene C–H as isoprene
1631	6130	C–H vinylidene C–H, associated with (CH₂=C<)	C–H, vinylidene (secondary alkyl group), CH₂=C<	Hydrocarbons, aliphatic
1635	6120	C–H vinyl C–H, associated with (CH₂=CH–)	C–H, vinyl (hexene), CH₂=CH–	Hydrocarbons, aliphatic
1637	6110	C–H from vinyl group as (CH₂=CH–)	C–H pendant vinyl group of the 1,2 unit for liquid carboxylated poly(acrylonitrile-co-butadiene) [or nitrile rubber (NBR)]	C–H from vinyl-associated group
1654	6045	C–H methyl C–H, nitro (CH₃NO₂)	C–H methyl, CH₃NO₂	Nitro (CH₃) as (CH₃NO₂)
1655	6025/6060 doublet	C–H methyl C–H, brominated (CH₃Br)	C–H methyl, CH₃Br	Halogenated (CH₃Br)
1661	6020	C–H methyl C–H, iodine (CH₃I)	C–H methyl, CH₃I	Halogenated (CH₃I)
1661	6000/6040 doublet	C–H methyl C–H, chlorinated CH₃	C–H methyl, CH₃Cl	Halogenated (CH₃Cl)
1664	6011	C–H methyl C–H, OH associated as (ROHCH₃)	C–H methyl, OH associated as ROHCH₃	Alcohols, diols
1671	5985 (5988)	C–H aromatic C–H (aryl)	CHv+CHv (12 + 1), benzene band assignment	C–H aryl
1678	5960	C–H methyl C–H, carbonyl adjacent as (.C=OCH₃)	C–H methyl, carbonyl adjacent as (.C=OCH₃)	Ketones
1680	5952	C–H aromatic (ArCH)	C–H (2v), aromatic C–H	Hydrocarbons, aromatic
1680	5952	Reference — classic filter instrument	CH – Aromatic used as classic reference filter for filter instruments	Reference (classic)
1682	5946	C–H methyl C–H, carbonyl associated as one C removed (.C=OCH₂CH₃)	C–H methyl, carbonyl associated as one C removed (.C=OCH₂CH₃)	Ketones
1685	5935	C–H aromatic C–H	C–H (2v), ArC–H	Hydrocarbons, aromatic
1688	5925	C–H methyl C–H, OH associated as (ROHCH₃)	C–H methyl, OH associated as ROHCH₃	Alcohols, diols

(continued)

Nanometers (nm)	Wavenumbers (cm⁻¹)	Functional Group	Spectra–Structure	Material Type
1689	5920 (5914)	C–H aromatic C–H (aryl)	CHv+CHv (15 + 5), benzene band assignment	C–H aryl
1690	5915–5925	CONH₂ specifically due to peptide N–H and C=O groups at right angles to the line of the peptide backbone referred to as the β-sheet structure	CONH₂ specifically due to peptide β-sheet structures	Proteins as normalized 2nd-derivative spectra of proteins in aqueous solution
1693	5905	C–H methyl C–H branched	C–H as RCH₃CHR branched	Aliphatic C–H
1693	5905	C–H methyl C–H terminal	C–H as R-CH₃ terminal	Aliphatic C–H
1693	5908	C–H methyl C–H, carbonyl associated as one C removed (.C=OCH₂CH₃)	C–H methyl, carbonyl associated as one C removed (.C=OCH₂CH₃)	Ketones
1694	5903	C–H methyl C–H, (.CH₃) (Asymmetric)	C–H (2v), methyl C–H (asymmetric)	Hydrocarbons, aliphatic
1694	5902	C–H methyl C–H, iodine (CH₃I)	C–H methyl, CH₃Br	Halogenated (CH₃I)
1695	5900	C–H methyl (.CH₃)	C–H (2v), .CH₃	Hydrocarbons, methyl
1695	5900 (gas phase)	C–H methyl C–H, brominated (CH₃Br)	C–H methyl, CH₃Br	Halogenated (CH₃Br)
1695	5898	C–H methyl C–H, carbonyl adjacent as (.C=OCH₃)	C–H methyl, carbonyl adjacent as (.C=OCH₃)	Ketones
1700	5882 (gas phase)	C–H methyl C–H, chlorinated (CH₃Cl)	C–H methyl, CH₃Cl	Halogenated (CH₃Cl)
1701	5880	C–H methyl C–H, ether associated as (R–O–CH₃)	C–H methyl, ether associated as R–O–CH₃	Ethers
1701	5880	C–H methyl C–H, OH associated as (ROHCH₃)	C–H methyl, OH associated as ROHCH₃	Alcohols, diols
1702	5874	C–H methyl C–H, nitro (CH₃NO₂)	C–H methyl, CH₃NO₂	Nitro (CH₃) as (CH₃NO₂)
1703	5872	C–H methyl C–H branched	C–H as RCH₃CHR branched	Aliphatic C–H
1704	5869	C–H methyl C–H, (.CH₃)	C–H (2v), methyl C–H (symmetric)	Hydrocarbons, aliphatic
1705	5865	C–H methyl (.CH₃)	C–H (2v), .CH₃	Hydrocarbons, methyl
1709	5853	C–H methyl C–H terminal	C–H as R–CH₃ terminal	Aliphatic C–H
1711	5845	C–H methyl C–H, iodine (CH₃I)	C–H methyl, CH₃Br	Halogenated (CH₃I)
1725	5797	C–H methylene (.CH₂)	C–H (2v), .CH₂	Hydrocarbons, methylene

1727	C–H methylene (.CH₂) (asymmetric) → C–H (2v), methylene; C–H (asymmetric)	Hydrocarbons, aliphatic
1732	C–H methyl C–H, OH associated as (ROHCH₃) → C–H methyl, OH associated as ROHCH₃	Alcohols, diols
1733	C–H methyl C–H, ether associated as (R–O–CH₃) → C–H methyl, ether associated as R–O–CH₃	Ethers
1735	C–H methyl C–H, amine associated as NH₂CH₃ → C–H methyl, amine associated as RN(CH₃)₂	Amines
1736	C–H methyl C–H, aromatic associated (ArCH₃) → C–H methyl, aromatic (ArCH₃)	Aromatic (ArCH₃)
1738	CONH₂ specifically due to C=O hydrogen bonded to the N–H of the peptide link termed the α-helix structure → CONH₂ specifically due to the α-helix peptide structure	Proteins as normalized 2nd-derivative spectra of proteins in aqueous solution
1740	S-H thiol (.S–H) → S–H (2v), S–H	Thiols
1744	C–H methyl C–H, aromatic associated (ArCH₃) → C–H methyl, aromatic (ArCH₃)	Aromatic (ArCH₃)
1748	C–H (2v) from silicone → C–H (2v) stretch from silicone (dimethyl siloxane)	Silicone (dimethyl siloxane)
1762	C–H methylene (.CH₂) (symmetric) → C–H (2v), methylene; C–H (symmetric)	Hydrocarbons, aliphatic
1765	C–H methylene (.CH₂) → C–H (2v), .CH₂	Hydrocarbons, methylene
1770	C–H methyl C–H, aromatic associated (ArCH₃) → C–H methyl, aromatic (ArCH₃)	Aromatic (ArCH₃)
1780	C–H methylene (.CH₂) → C–H (2v), .CH₂	Cellulose
1790	O–H from water → O–H combination	Water
1820	O–H/C–H combination → O–H stretching and C–O stretching (3vₛ) combination	Cellulose
1860	C–Cl chlorinated organics (.C–Cl group) → C–Cl (7v), .C–Cl	Chlorinated hydrocarbons
1892	O–H hydrogen bonding between water and exposed polyvinyl alcohol OH → O–H assigned as a one-to-one hydrogen bonded to an isolated OH in the manner as OH::OH₂ from the effect of the hydration of the isolated alcohol OH — this pertains to the interactions of water with the OH groups in poly(ethylene-co-vinyl alcohol) (EVOH)	Water and polyvinyl alcohol OH
1900	C=O carbonyl (.C=O) → C=O (3v), .C=OOH	Acids, carboxylic

(continued)

Nanometers (nm)	Wavenumbers (cm⁻¹)	Functional Group	Spectra–Structure	Material Type
1908	5241	P–OH phosphate (.P–OH)	O–H (2v). P–OH	Phosphate
1920	5208	C=O amide (.C=ONH)	C=O (3v), C=ONH	Amide
1923	5200	O–H assigned to molecular water [O–H (.O–H and HOH)]	O–H assigned to molecular water (O–H stretching and HOH deformation combination)	O–H molecular water
1928	5186	O–H (.O–H and HOH)	O–H stretching and HOH deformation combination from water molecules in the 3-aminopropyltriethoxysilane-ethanol-water system	3-aminopropyltriethoxy-silane-ethanol-water system
1930	5181	O–H (.O–H and HOH)	O–H stretching and HOH bending combination	Polysaccharides
1933	5173	Si–O–H stretch + Si–O–Si combination from silicone	Si–O–H stretch + Si–O–Si deformation combination from silicone (dimethyl siloxane)	Silicone (dimethyl siloxane)
1940	5155	OH — classic filter instrument	Classic filter instrument	Moisture (classic)
1940	5155	O–H (H–O–H) water	O–H stretching and HOH bending combination	Water
1942	5150	O–H hydrogen bonding between water and exposed polyvinyl alcohol OH	O–H interaction of multiple hydrogen atoms from poly(ethylene-co-vinyl alcohol) (EVOH) bonded to surrounding associated OH groups (from water) without clustering	Water and polyvinyl alcohol OH
1950	5128	C=O esters and acids (.C=OOR)	C=O (3v), C=OOR	Acids and esters
1960	5100	N–H combination band from primary amides (R–C=O–NH$_2$)	N–H [vN–H asymmetric and Amide II deformation (N–H in-plane bending) combination] for primary amides	N–H combination band from primary amides
1960	5102	O–H polymeric (.O–H)	O–H stretching and HOH bending combination	Polysaccharides

5095	N–H combination, primary aromatic amine (*m*-NO$_2$)	N–H (vN–H and δN–H combination), primary aromatic amine in CCl$_4$ as *meta*-NO$_2$ substituent	Aromatic amine
5090	O–H and CH combination from methanol	vO–H and δC–H combination from CH$_3$OH	Methanol
5084	N–H combination, primary aromatic amine (*o*-NO$_2$)	N–H (vN–H and δN–H combination), primary aromatic amine in CCl$_4$ as *ortho*-NO$_2$ substituent	Aromatic amine
5082	N–H combination, primary aromatic amine (*m*-Cl)	N–H (vN–H and δN–H combination), primary aromatic amine in CCl$_4$ as *meta*-Cl substituent	Aromatic amine
5075	N–H combination, primary aromatic amine (*p*-Cl)	N–H (vN–H and δN–H combination), primary aromatic amine in CCl$_4$ as *para*-Cl substituent	Aromatic amine
5072	N–H combination, primary aromatic amine (no other substituents)	N–H (vN–H and δN–H combination), primary aromatic amine in CCl$_4$, no other substituents	Aromatic amine
5072	N–H combination, primary aromatic amine (*o*-Cl)	N–H (vN–H and δN–H combination), primary aromatic amine in CCl$_4$ as *ortho*-Cl grouping	Aromatic amine
5071	N–H combination, from primary and secondary aromatic amines (mixtures of aniline and N-ethylaniline in CCl$_4$)	N–H (vN–H and δN–H combination), primary aromatic amine in CCl$_4$ as *para*-Cl substituent from primary and secondary aromatic amines	Primary and secondary aromatic amines
5062	N–H combination, primary aromatic amine (*p*-CH$_3$)	N–H (vN–H and δN–H combination), primary aromatic amine in CCl$_4$ as *para*-CH$_3$ grouping	Aromatic amine
5057	N–H combination, primary aromatic amine (*o*-OCH$_3$)	N–H (vN–H and δN–H combination), primary aromatic amine in CCl$_4$ as *ortho*-OCH$_3$ substituent	Aromatic amine
5051	N–H amide II (.CONH$_2$)	N–H stretching (asymmetric) and N–H in-plane bending combination	Amides/proteins

1963.0			
1965			
1967.0			
1968.0			
1970.5			
1971.5			
1971.5			
1972			
1975.5			
1977.5			
1980			

(continued)

Nanometers (nm)	Wavenumbers (cm^{-1})	Functional Group	Spectra–Structure	Material Type
1980.5	5049	N–H combination, primary aromatic amine (p-NH$_2$)	N–H (νN–H and δN–H combination), primary aromatic amine in CCl$_4$ as para-NH$_2$ grouping	Aromatic amine
1990	5025	N–H amide: NH$_2$–C=ONH$_2$	N–H stretching and N–H bending combination	Urea
1990	5025	N–H/C–N combination band from urea (NH$_2$–C=O–NH$_2$)	N–H [νN–H asymmetric and amide III deformation (C–N stretching/N–H in-plane bending) combination] for primary amides	Urea
2000	5000	N–H ammonia in water	N–H (νN–H and δN–H combination) for NH$_3$ (ammonia) in water	Ammonia
2010	4975	N–H/C–N combination band from primary amides (R–C=O–NH$_2$)	N–H [νN–H symmetric and amide III deformation (C–N stretching/N–H in-plane bending) combination] for primary amides	Primary amides
2012	4970	N–H/C=O combination from polyamide 11	N–H/C=O [bonded NH stretching and amide I (2νC=O stretching) combination] from polyamide 11	Polyamide 11
2016	4960	O–H stretching and bending combination from methanol	νO–H and δO–H combination from CH$_3$OH	Methanol O–H
2024	4940	N–H/C=O combination from of native RNase A	N–H/C=O [bonded NH stretching and amide I (2νC=O stretching) combination] of native RNase A	RNase A
2030	4926	C=O amide: NH$_2$–C=ONH$_2$	C=O (3ν), C=ONH$_2$	Urea
2030	4925	N–H/C–N combination band from primary amides (R–C=O–NH$_2$)	N–H [νN–H asymmetric and Amide III deformation (C–N stretching/N–H in-plane bending) combination] for primary amides	Primary amides

2040	N–H combination band from urea (NH$_2$–C=O–NH$_2$)	4902	N–H [vN–H symmetric and Amide II deformation (N–H in-plane bending) combination] for primary amides	Urea
2050	N–H combination band (RNase A) – C=O amide I band	4878	N–H native RNase A combination band at 4867 cm^{-1} shifting due to thermal unfolding (C=O amide I band)	N–H from protein
2050	N–H/C=O amide as (.CONH.) and (.CONH$_2$) from native RNase A (thermal unfolding)	4878	N–H stretching and C=O stretching (amide I) combination band observed in the thermal unfolding observed in native RNase A	RNase A
2050	N–H/C–N/N–H amide II and amide III combination (.CONH.) and .CONH$_2$)	4878	N–H in-plane bend and C–N stretching and N–H in-plane bend combination	Amides/proteins
2053	N–H combination from polyamide 11	4870	Bonded NH stretching and amide II (N–H in-plane bending) from polyamide 11	Polyamide 11
2053	N–H band found at 4867 cm^{-1} for native RNase A shifting to 4878 cm^{-1} upon thermal unfolding	4867–4878	N–H from CONH$_2$ as thermal unfolding of RNase A protein in aqueous solution (assigned to an N–H combination band)	Protein thermal unfolding
2055	CONH$_2$ specifically due to peptide N–H and C=O groups at right angles to the line of the peptide backbone referred to as the β-sheet structure	4865	CONH$_2$ specifically due to peptide β-sheet structures	Proteins as normalized 2nd-derivative spectra of proteins in aqueous solution
2055	N–H combination band (RNase A) – C=O amide I band	4867	N–H combination band found in the spectrum of native RNase A (C=O amide I band)	N–H from protein
2055	N–H from gamma-valerolactam	4865	N–H (vN–H and δN–H combination) for Gamma-valerolactam	Gamma-valerolactam
2055	N–H/C=O amide as (.CONH.) and (.CONH$_2$)	4866	N–H stretching and C=O stretching (amide I) combination	Amides/proteins

(continued)

Nanometers (nm)	Wavenumbers (cm^{-1})	Functional Group	Spectra–Structure	Material Type
2055	4867	N–H/C=O amide as (.CONH .) and .CONH$_2$) from native RNase A	N–H stretching and C=O stretching (amide I) combination band in the spectrum of native RNase A	RNase A
2060	4855	HN...O=C band of amide A	CONH$_2$ combination of amide A and amide II	Proteins (polypeptides)
2060	4854	N–H amide as (.CONH .) and (.CONH$_2$)	N–H (3δ) and N–H stretching combination	Amides/proteins
2060	4850	N–H combination band from secondary amides in proteins	N–H [νN–H and amide II deformation (N–H in-plane bending) combination] for secondary amides in proteins	N–H from protein
2070	4831	N–H amide: (N–H deformation)	N–H (δ)	Urea
2075	4820	N–H combination band from secondary amides in native RNase A	N–H [νN–H and amide II deformation (N–H in-plane bending) combination] for secondary amides in native RNase A	Native RNase A
2080	4808	N–H/C–N combination band from urea (NH$_2$–C=O–NH$_2$)	N–H/C–N [νN–H asymmetric and amide III deformation (C–N stretching/N–H in-plane bending) combination] for urea	Urea
2080	4808	O–H and C–O stretching and bending combination from methanol	vO–H and δO–H combination from CH$_3$OH	Methanol
2080	5550–4550 (broad)	O–H broad band occurring in polyols, alcohols, water, ethylene vinyl alcohols and copolymers containing O–H groups	O–H broad band, due to vO–H and δO–H combination	Polyols, alcohols, water
2083	4800	O–H related combination from water change in phase and N–H/C–N combination band from urea (NH$_2$–C=O–NH$_2$) from ovalbumin	O–H related combination from water change in phase with the increase in protein concentration at pH above 2.8 overlapping with a band representing the N–H/C–N [ν$_S$N–H asymmetric and amide III deformation (C–N stretching/N–H in-plane bending) combination]	Ovalbumin protein

2090		N–H from gamma-valerolactam	N–H [NH stretching and amide II deformation (N–H in-plane bending) combination] for gamma-valerolactam	Gamma-valerolactam
2090		O–H polymeric (.O–H)	O–H combination	Polymeric .OH
2096		O–H deformation band, alcohols or water	3δ O–H of mid-infrared band occurring near 1420 cm^{-1}	Alcohols or water O–H
2100		CHO — classic filter instrument	Classic filter instrument	Carbohydrate
2100		C=O–O polymeric (C=O and C–O stretching)	C=O–O (4v)	Polysaccharides
2100		O/H/C–O polymeric (.O–H and .C–O)	O–H bending and C–O stretching combination	Polysaccharides
2120		N–H/C–N from gamma-valerolactam	N–H/C–N [vN–H and amide III deformation (C–N stretching/N–H in-plane bending) combination] for gamma-valerolactam	Gamma-valerolactam
2123		O–H/C–O stretching combination from methanol	vO–H and vC–O combination from CH$_3$OH	Methanol O–H
2127		N–H/C=O combination from polyamide 11	N–H/C=O [2 × amide II (N–H in-plane bending) and amide I (2vC=O stretching) combination] from polyamide 11	Polyamide 11
2140		C–H/C=O lipid associated (.RC=CH and RC=O)	C–H stretching and C=O stretching combination and C–H deformation combination	Lipids
2145		N–H/C–N/C=O from gamma-valerolactam	N–H/C–N/C=O [2 × amide I (2vC=O stretching) and amide III deformation (C–N stretching/N–H in-plane bending) combination] for gamma-valerolactam	Gamma-valerolactam
2148 (4675)	4655	C–H aromatic C–H (aryl)	CHv+CCv (15 + 9), benzene band assignment	C–H aryl
2154 (4644)	4642	C–H aromatic C–H (aryl)	CHv+CCv (12 + 16), benzene band assignment	C–H aryl
2167 (4625)	4615	C–H aromatic C–H (aryl)	CCv+CHv (16 + 5), benzene band assignment	C–H aryl

(continued)

Nanometers (nm)	Wavenumbers (cm⁻¹)	Functional Group	Spectra–Structure	Material Type
2167	4615	CONH$_2$, specifically due to C=O hydrogen bonded to the N–H of the peptide link termed the α-helix structure	CONH$_2$, specifically due to the α-helix peptide structure	Proteins as normalized 2nd-derivative spectra of proteins in aqueous solution
2170	4608	C–H alkenes (.HC=CH)	C–H stretching and C–H deformation combination	Alkenes
2174	4600	CONH$_2$ as combination of amide B and amide II modes (amide B/amide II) from ovalbumin	CONH$_2$, as combination of amide B and amide II modes (amide B/amide II)	Amides
2174	4600	CONH$_2$ specifically due to amide B and amide II modes (amide B/II) from ovalbumin protein side chains seen at shifting to lower wavenumber with considerable broadening at lowered pH (from pH 5.0 to 2.4)	CONH$_2$ specifically due to amide B and amide II modes (amide B/II)	Proteins at low pH
2180	4587	N–H — classic filter instrument	N–H classic filter instrument	Protein
2180	4587	N–H proteins: N–H (3ν$_B$)	N–H (3δ)	Proteins/amino acids
2180	4590	N–H/C–N/C=O combination band from secondary amides in proteins	N–H/C–N/C=O [2 × amide I (2νC=O stretching) and amide III deformation (C–N stretching/N–H in-plane bending) combination] for secondary amides in proteins	Protein
2180	4687	N–H/C–N/C=O combination band from urea (NH$_2$–C=O–NH$_2$)	N–H/C–N/C=O [2 × amide I (2νC=O stretching) and amide III deformation (C–N stretching/N–H in-plane bending) combination] for urea	Urea
2183	4586	N–H/C–N/N–H combination from polyamide 11	N–H/C–N/N–H [bonded NH stretching and amide III (C–N stretching/N–H in-plane bending) combination] from polyamide 11	Polyamide 11
2188	4570 (4584)	C–H aromatic C–H (aryl)	CCν+CHν (13 + 1), benzene band assignment	C–H aryl
2200	4545	CHO carbohydrate (.CHO)	C–H stretching and C=O combination	Carbohydrates

Wavenumber	Assignment	Description	Material
2205	N–H primary amine band of diamino- compound	N–H (3ν) band of primary amine as bisphenol A (I) resins cured with 4,4'-diaminodiphenyl sulfone (II) hardener	Amine, primary
2206	C–H aromatic C–H (aryl)	CCν+CHν (13 + 15), benzene band assignment	C–H aryl
2207	CONH$_2$ specifically due to peptide N–H and C=O groups at right angles to the line of the peptide backbone referred to as the β-sheet structure	CONH$_2$ specifically due to peptide β-sheet structures	Proteins as normalized 2nd-derivative spectra of proteins in aqueous solution
2210	N–H ammonia in water	N–H (3ν) for NH$_3$ (ammonia) in water	Ammonia in water
2212	C=O/C–N/N–H combination from polyamide 11	C=O/C–N/N–H [2 × amide I (2νC=O stretching) and amide III (C–N stretching/N–H in-plane bending) combination] from polyamide 11	Polyamide 11
2220	N–H combination band from urea (NH$_2$–C=O–NH$_2$)	N–H (νN–H asymmetric and NH$_2$ rocking) combination	N–H from urea
2230	CHO — classic filter instrument	Classic filter instrument	Reference (classic)
2270	CHO — classic filter instrument	Classic filter instrument	Lignin
2270	CONH$_2$ specifically due to peptide N–H and C=O groups at right angles to the line of the peptide backbone referred to as the β-sheet structure	CONH$_2$ specifically due to peptide β-sheet structures	Proteins as normalized 2nd-derivative spectra of proteins in aqueous solution
2270	O–H/C–H cellulose (.OH and .C–O)	O–H stretching and C–O stretching combination	Cellulose
2273	O–H/C–O from glucose	O–H/C–O glucose absorption from O–H stretching and C–O stretching combination	Glucose
2280	C–H starch (.C–H and CH$_2$)	C–H stretching and CH$_2$ deformation	Polysaccharides
2290	CONH$_2$ specifically due to C=O hydrogen bonded to the N–H of the peptide link termed the α-helix structure	CONH$_2$ specifically due to C=O hydrogen bonded to the N–H of the α-helix peptide structure	Proteins as normalized 2nd-derivative spectra of proteins in aqueous solution

(continued)

Nanometers (nm)	Wavenumbers (cm⁻¹)	Functional Group	Spectra–Structure	Material Type
2294	4360 (4379)	C–H aromatic C–H (aryl)	CHv+CHδ (12 + 3), benzene band assignment	C–H aryl
2295		C–H (3δ) from silicone	C–H (3δ) bend from silicone (dimethyl siloxane)	C–H from silicone (dimethyl siloxane)
2300		C–H (.C–H bending)	C–H (3δ)	Amides
2308		C–H methylene C–H, associated with linear aliphatic R(CH₂)ₙR	C–H (2vCH₂, asymmetric stretching and δCH₂) combination	Hydrocarbons, aliphatic
2310		C–H (.C–H bending)	C–H (3δ)	Lipids
2310		CHO — classic filter instrument	Classic filter instrument	Lipids/oils
2318		C–H methylene C–H, associated with branched aliphatic RC(CH₃)₃ or RCH(CH₃)₂	C–H (2vCH₂ asymmetric stretching and δCH₂) combination	Hydrocarbons, aliphatic
2322		C–H (.C–H and CH₂)	C–H stretching and CH₂ deformation combination	Polysaccharides
2330		C–H (.C–H and CH₂)	C–H stretching and CH₂ deformation combination	Polysaccharides
2335		C–H (.C–H and CH₂)	C–H stretching and CH₂ deformation combination	Polysaccharides
2336		CHO — classic filter instrument	Classic filter instrument	Cellulose
2345		C–H methylene C–H, associated with ovalbumin protein side chains seen at pH 5.0	C–H (2vCH₂ symmetric stretching and δCH₂) combination band from ovalbumin protein side chains seen at pH 5.0	C–H from ovalbumin protein side chains
2347		C–H methylene C–H, associated with linear aliphatic R(CH₂)ₙR	C–H (2vCH₂ symmetric stretching and δCH₂) combination	Hydrocarbons, aliphatic
2352		C–H (.C–H bending)	C–H (3δ)	Polysaccharides
2352	(4263)	C–H aromatic C–H (aryl)	CHv+CHδ (12 + 17), benzene band assignment	C–H aryl
2363		C–H methylene C–H, associated with branched aliphatic RC(CH₃)₃ or RCH(CH₃)₂	C–H (2vCH₂, symmetric stretching and δCH₂) combination	Hydrocarbons, aliphatic
2380		C–H/C–C (.C–H and .C-C)	C–H stretching and C-C stretching combination	Lipids
2387	4190 (4198)	C–H aromatic C–H (aryl)	CCδ+CHv (17 + 5), benzene band assignment	C–H aryl

nm	cm⁻¹		Band assignment	
2407	4155 (4175)	C–H aromatic C–H (aryl)	CHv+CHδ (15 + 10), benzene band assignment	C–H aryl
2440	4099	C–H aromatic C–H (aryl)	CCδ+CHv (14 + 1), benzene band assignment	C–H aryl
2444	4091	C–H aromatic C–H (aryl)	CHv+CCv (12 + 2), benzene band assignment	C–H aryl
2445	4090	CONH₂, specifically due to C=O hydrogen bonded to the N–H of the peptide link termed the α-helix structure	CONH₂, specifically due to the α-helix peptide structure	Proteins as normalized 2nd-derivative spectra of proteins in aqueous solution
2458	4068	C–H methyl C–H, associated with linear aliphatic CH₃ (CH₂)ₙ CH₃	C–H (3δ CH₃)	Hydrocarbons, aliphatic
2463	4060	CONH₂ specifically due to peptide N–H and C=O groups at right angles to the line of the peptide backbone referred to as the β-sheet structure	CONH₂, specifically due to peptide β-sheet structures	Proteins as normalized 2nd-derivative spectra of proteins in aqueous solution
2469	4050 (4060)	C–H aromatic C–H (aryl)	CCδ+CHv (14 + 15), benzene band assignment	C–H aryl
2470	4049	C–H (.CH₂)	C–H combination	Lipids, aliphatic compounds
2470	4049	C–H methyl C–H, associated with branched aliphatic RC(CH₃)₃ or RCH(CH₃)₂	C–H (3δ CH₃)	Hydrocarbons, aliphatic
2470	4049	C–N–C amide: (.C–N–C)	C–N–C (2v)	Proteins
2477	4037	C–H methyl C–H, associated with aromatic (ArCH₃)	C–H (3δ CH₃)	Hydrocarbons, aromatic
2488	4019	C–H/C–C (.C–H and .C–C)	C–H stretching and C–C stretching combination	Cellulose
2500	4000	C–H/C–C/C–O–C (.C–H and .C–C and .C–O–C)	C–H stretching and C–C and C–O–C stretching combination	Polysaccharides
2513	3980	C–H aromatic C–H (aryl)	CHv+CCδ (15 + 6), benzene band assignment	C–H aryl
2525	3960	C–H aromatic C–H (aryl)	CCω+CHv (19 + 15), benzene band assignment	C–H aryl
2530	3953	C–N–C amide: (.C–N–C)	C–N–C (2v) asymmetric	Amide
2540	3937	C–H aromatic C–H (aryl)	CCω+CHv (11+ 12), benzene band assignment	C–H aryl

(continued)

Nanometers (nm)	Wavenumbers (cm^{-1})	Functional Group	Spectra–Structure	Material Type
2679	3733	C–H aromatic C–H (aryl)	CHo+CHv (4 + 1), benzene band assignment	C–H aryl
2687	3722	C–H aromatic C–H (aryl)	CCo+CHv (7 + 5), benzene band assignment	C–H aryl
2708	3693	C–H aromatic C–H (aryl)	CCo+CHv (8 + 5), benzene band assignment	C–H aryl
2740	3650	O–H from primary alcohols as (–CH–OH)	O–H (v) (–CH$_2$–OH), primary alcohols	Primary alcohols
2747	3640	C–H aromatic C–H (aryl)	CHv+CCδ (12 + 18), benzene band assignment	C–H aryl
2762	3620	O–H from secondary alcohols as (–CH–OH)	O–H (v) (–CH–OH), secondary alcohols	Secondary alcohols
2768	3613	C–H aromatic C–H (aryl)	CCo+CHv (18 + 5), benzene band assignment	C–H aryl
2770	3610	O–H from tertiary alcohols as (–C–OH)	O–H (v) (–C–OH), tertiary alcohols	Tertiary alcohols
3030	3300	CONH$_2$ (HN...O=C band) of amide A	CONH$_2$ as amide A for the polypeptides	Amide A from proteins (polypeptides)
3049	3280	O–H stretching vibration (v) of water molecules in bound-water hydrogen-bonding in turquoise minerals	O–H (v) (as intense and very broad band assigned to bound-water hydrogen-bonding in turquoise minerals from Arizona and Senegal with a formula of Cu(Al6-x,Fex)(PO4)4(OH)8.4H2O	Minerals
3067	3260	O–H stretching vibration (v) of water molecules in protein-water hydrogen bonding	O–H (v) (as intense and very broad band assigned to protein-water hydrogen bonding	Protein
3521	2840	O–H stretching vibrations of water molecules in water-water hydrogen bonding	O–H as intense and very broad band assigned to water-water hydrogen bonding	Water

ABBREVIATIONS AND SYMBOLS

(1ν)   *Fundamental* stretching vibrational absorption band
(2ν)   *First* overtone of fundamental stretching band
(3ν)   *Second* overtone of fundamental stretching band
(4ν)   *Third* overtone of fundamental stretching band
(5ν)   *Fourth* overtone of fundamental stretching band
(6ν)   *Fifth* overtone of fundamental stretching band

(1δ)   *Fundamental* bending vibrational absorption band
(2δ)   *First* overtone of fundamental bending band
(3δ)   *Second* overtone of fundamental bending band
(4δ)   *Third* overtone of fundamental bending band

(ω)   Deformation (rocking or wagging)

* Data for this appendix is referenced within the chapters of this book. Additional limited data is reprinted from Burns, Don and Ciurczak, Emil (eds.), *Handbook of Near Infrared Analysis*, 2nd ed., Marcel Dekker, 2001, pp. 431–433. Used with permission.

Appendix 4b: Spectra–Structure Correlations for Near-Infrared (in Ascending Alphabetical Functional Group Order)*

---

* Data for this appendix is referenced within the chapters of this book. Additional limited data is reprinted from Burns, Don and Ciurczak, Emil (eds.), *Handbook of Near Infrared Analysis*, 2nd ed., Marcel Dekker, 2001, pp. 431–433. Used with permission.

Functional Group	Nanometers (nm)	Wavenumbers (cm⁻¹)	Spectra-Structure	Material Type
C–Cl chlorinated organics (.C–Cl group)	1860	5376	C–Cl (7v), .C–Cl	Chlorinated hydrocarbons
C–H (.C–H and CH₂)	2322	4307	C–H stretching and CH₂ deformation combination	Polysaccharides
C–H (.C–H and CH₂)	2330	4292	C–H stretching and CH₂ deformation combination	Polysaccharides
C–H (.C–H and CH₂)	2335	4283	C–H stretching and CH₂ deformation combination	Polysaccharides
C–H (.C–H bending)	2300	4348	C–H (3δ)	Amides
C–H (.C–H bending)	2310	4329	C–H (3δ)	Lipids
C–H (.C–H bending)	2352	4252	C–H (3δ)	Polysaccharides
C–H (.CH₂)	2470	4049	C–H combination	Lipids, aliphatic compounds
C–H (2v) from silicone	1748	5721	C–H (2v) stretch from silicone (dimethyl siloxane)	C–H from silicone (dimethyl siloxane)
C–H (3δ) from silicone	2295	4357	C–H (3δ) bend from silicone (dimethyl siloxane)	C–H from silicone (dimethyl siloxane)
C–H (aromatic C–H)	1143	8749	C–H (3v), Ar.C–H	Hydrocarbons, aromatic
C–H acrylate C–H as (CH₂=CHCOO⁻)	1621	6169	C–H, combination of two stretching modes	Acrylate
C–H alkene (.HC=CH)	1170	8547	C–H (3v), HC=CH	Alkenes, polyenes
C–H alkene, .=CH₂	1620	6173	C–H (2v), =CH₂	Alkenes
C–H alkenes (.HC=CH)	2170	4608	C–H stretching and C–H deformation combination	Alkenes
C–H aromatic (ArCH)	714	13,986	C–H (5v), aromatic C–H	Hydrocarbons, aromatic
C–H aromatic (ArCH)	876	11,655	C–H (4v), aromatic C–H	Hydrocarbons, aromatic
C–H aromatic (ArCH)	1142	8754	C–H (3v), aromatic C–H	Hydrocarbons, aromatic
C–H aromatic (ArCH)	1680	5952	C–H (2v), aromatic C–H	Hydrocarbons, aromatic
C–H aromatic (ArC–H)	1446	6916	C–H combination, ArC–H	Hydrocarbons, aromatic
C–H aromatic C–H	1417	7057	C–H combination, Ar.C–H	Hydrocarbons, aromatic
C–H aromatic C–H	1685	5935	C–H (2v), ArC–H	Hydrocarbons, aromatic
C–H aromatic C–H (aryl)	1671	5985 (5988)	CHv+CHv (12 + 1), benzene band assignment	C–H aryl
C–H aromatic C–H (aryl)	1689	5920 (5914)	CHv+CHv (15 + 5), benzene band assignment	C–H aryl
C–H aromatic C–H (aryl)	2148	4655 (4675)	CHv+CCv (15 + 9), benzene band assignment	C–H aryl
C–H aromatic C–H (aryl)	2154	4642 (4644)	CHv+CCv (12 + 16), benzene band assignment	C–H aryl
C–H aromatic C–H (aryl)	2167	4615 (4625)	CCv+CHv (16 + 5), benzene band assignment	C–H aryl

C–H aromatic C–H (aryl)	2188	CCv+CHv (13 + 1), benzene band assignment	4570 (4584)	C–H aryl
C–H aromatic C–H (aryl)	2206	CCv+CHv (13 + 15), benzene band assignment	4532 (4549)	C–H aryl
C–H aromatic C–H (aryl)	2294	CHv+CHδ (12 + 3), benzene band assignment	4360 (4379)	C–H aryl
C–H aromatic C–H (aryl)	2352	CHv+CHδ (12 + 17), benzene band assignment	4252 (4263)	C–H aryl
C–H aromatic C–H (aryl)	2387	CCδ+CHv (17 + 5), benzene band assignment	4190 (4198)	C–H aryl
C–H aromatic C–H (aryl)	2407	CHv+CHδ (15 + 10), benzene band assignment	4155 (4175)	C–H aryl
C–H aromatic C–H (aryl)	2440	CCδ+CHv (14 + 1), benzene band assignment	4099	C–H aryl
C–H aromatic C–H (aryl)	2444	CHv+CCv (12 + 2), benzene band assignment	4091	C–H aryl
C–H aromatic C–H (aryl)	2469	CCδ+CHv (14 + 15), benzene band assignment	4050 (4060)	C–H aryl
C–H aromatic C–H (aryl)	2513	CHv+CCδ (15 + 6), benzene band assignment	3980 (3986)	C–H aryl
C–H aromatic C–H (aryl)	2525	CCo+CHv (19 + 15), benzene band assignment	3960 (3958)	C–H aryl
C–H aromatic C–H (aryl)	2540	CCo+CHv (11 + 12), benzene band assignment	3937 (3935)	C–H aryl
C–H aromatic C–H (aryl)	2679	CHo+CHv (4 + 1), benzene band assignment	3733	C–H aryl
C–H aromatic C–H (aryl)	2687	CCo+CHv (7 + 5), benzene band assignment	3722 (3735)	C–H aryl
C–H aromatic C–H (aryl)	2708	CCo+CHv (8 + 5), benzene band assignment	3693 (3697)	C–H aryl
C–H aromatic C–H (aryl)	2747	CHv+CCδ (12 +18), benzene band assignment	3640 (3643)	C–H aryl
C–H aromatic C–H (aryl)	2768	CCo+CHv (18 + 5), benzene band assignment	3613 (3613)	C–H aryl
C–H from vinyl group as (CH₂=CH–)	1637	C–H pendant vinyl group of the 1,2 unit for liquid carboxylated poly(acrylonitrile-co-butadiene) [or nitrile rubber (NBR)]	6110	C–H from vinyl-associated group
C–H methyl (.CH₃)	908	C–H (4v), .CH₃	11,052–10,953	Hydrocarbons, methyl
C–H methyl (.CH₃)	1195	C–H (3v), .C–H₃	8368	Hydrocarbons, aliphatic
C–H methyl (.CH₃)	1360	C–H combination, .C–H₃	7353	Hydrocarbons, aliphatic
C–H methyl (.CH₃)	1695	C–H (2v), .CH₃	5900	Hydrocarbons, methyl
C–H methyl (.CH₃)	1705	C–H (2v), .CH₃	5865	Hydrocarbons, methyl
C–H methyl C–H branched	1693	C–H as RCH₂CHR branched	5905	Aliphatic C–H
C–H methyl C–H branched	1703	C–H as RCH₂CHR branched	5872	Aliphatic C–H
C–H methyl C–H terminal	1693	C–H as R–CH₃ terminal	5905	Aliphatic C–H
C–H methyl C–H terminal	1709	C–H as R–CH₃ terminal	5853	Aliphatic C–H
C–H methyl C–H, (.CH₃)	747	C–H (5v), methyl C–H	13,387	Hydrocarbons, aliphatic
C–H methyl C–H, (.CH₃)	915	C–H (4v), methyl C–H	10,929	Hydrocarbons, aliphatic
C–H methyl C–H, (.CH₃)	1194	C–H (3v), methyl C–H	8375	Hydrocarbons, aliphatic

(continued)

Functional Group	Nanometers (nm)	Wavenumbers (cm^{-1})	Spectra-Structure	Material Type
C–H methyl C–H, (,CH$_3$) (Asymmetric)	1694	5903	C–H (2v), methyl C–H (asymmetric)	Hydrocarbons, aliphatic
C–H methyl C–H, (,CH$_3$) (Symmetric)	1704	5869	C–H (2v), methyl C–H (symmetric)	Hydrocarbons, aliphatic
C–H methyl C–H, amine associated as NH$_2$CH$_3$	1735	5765	C–H methyl, amine associated as RN(CH$_3$)$_2$	Amines
C–H methyl C–H, aromatic associated (ArCH$_3$)	1736	5760	C–H methyl, aromatic (ArCH$_3$)	Aromatic (ArCH$_3$)
C–H methyl C–H, aromatic associated (ArCH$_3$)	1744	5735	C–H methyl, aromatic (ArCH$_3$)	Aromatic (ArCH$_3$)
C–H methyl C–H, aromatic associated (ArCH$_3$)	1770	5650	C–H methyl, aromatic (ArCH$_3$)	Aromatic (ArCH$_3$)
C–H methyl C–H, associated with aromatic (ArCH$_3$)	803	12,453	C–H (4vCH$_3$ and δCH$_3$) combination	Hydrocarbons, aromatic
C–H methyl C–H, associated with aromatic (ArCH$_3$)	1021	9796	C–H (3vCH$_3$ and δCH$_3$) combination	Hydrocarbons, aromatic
C–H methyl C–H, associated with aromatic (ArCH$_3$)	1370	7300	C–H (2vCH$_3$ and δCH$_3$) combination	Hydrocarbons, aromatic
C–H methyl C–H, associated with aromatic (ArCH$_3$)	2477	4037	C–H (3δ CH$_3$)	Hydrocarbons, aromatic
C–H methyl C–H, associated with branched aliphatic RC(CH$_3$)$_3$ or RCH(CH$_3$)$_2$	813	12,300	C–H (4vCH$_3$ and δCH$_3$) combination	Hydrocarbons, aliphatic
C–H methyl C–H, associated with branched aliphatic RC(CH$_3$)$_3$ or RCH(CH$_3$)$_2$	1021	9794	C–H (3vCH$_3$ and δCH$_3$) combination	Hydrocarbons, aliphatic
C–H methyl C–H, associated with branched aliphatic RC(CH$_3$)$_3$ or RCH(CH$_3$)$_2$	1396	7163	C–H (2vCH$_3$ and δCH$_3$) combination	Hydrocarbons, aliphatic
C–H methyl C–H, associated with branched aliphatic RC(CH$_3$)$_3$ or RCH(CH$_3$)$_2$	2470	4049	C–H (3δ CH$_3$)	Hydrocarbons, aliphatic

(continued)

C–H methyl C–H, associated with linear aliphatic CH_3 $(CH_2)_N$ CH_3	796	12,563	C–H ($4\nu CH_3$ and δCH_3) combination	Hydrocarbons, aliphatic
C–H methyl C–H, associated with linear aliphatic CH_3 $(CH_2)_N$ CH_3	1021	9795	C–H ($3\nu CH_3$ and δCH_3) combination	Hydrocarbons, aliphatic
C–H methyl C–H, associated with linear aliphatic CH_3 $(CH_2)_N$ CH_3	1390	7194	C–H ($2\nu_s CH_3$ and δCH_3) combination	Hydrocarbons, aliphatic
C–H methyl C–H, associated with linear aliphatic CH_3 $(CH_2)_N$ CH_3	2458	4068	C–H ($3\delta\ CH_3$)	Hydrocarbons, aliphatic
C–H methyl C–H, brominated (CH_3Br)	1650/1660	6060/6025 doublet	C–H methyl, CH_3Br	Halogenated (CH_3Br)
C–H methyl C–H, brominated (CH_3Br)	1695	5900 (gas phase)	C–H methyl, CH_3Br	Halogenated (CH_3Br)
C–H methyl C–H, carbonyl adjacent as (.C=OCH_3)	1678	5960	C–H methyl, carbonyl adjacent as (.C=OCH_3)	Ketones
C–H methyl C–H, carbonyl adjacent as (.C=OCH_3)	1695	5898	C–H methyl, carbonyl adjacent as (.C=OCH_3)	Ketones
C–H methyl C–H, carbonyl associated as one C removed (.C=OCH_2CH_3)	1682	5946	C–H methyl, carbonyl associated as one C removed (.C=OCH_2CH_3)	Ketones
C–H methyl C–H, carbonyl associated as one C removed (.C=OCH_2CH_3)	1693	5908	C–H methyl, carbonyl associated as one C removed (.C=OCH_2CH_3)	Ketones
C–H methyl C–H, chlorinated (CH_3Cl)	1700	5882 (gas phase)	C–H methyl, CH_3Cl	Halogenated (CH_3Cl)
C–H methyl C–H, chlorinated CH_3	1656/1667	6040/6000 doublet	C–H methyl, CH_3Cl	Halogenated (CH_3Cl)
C–H methyl C–H, ether associated as (R–O–CH_3)	1701	5880	C–H methyl, ether associated as R–O–CH_3	Ethers
C–H methyl C–H, ether associated as (R–O–CH_3)	1733	5770	C–H methyl, ether associated as R–O–CH_3	Ethers
C–H methyl C–H, iodine (CH_3I)	1661	6020	C–H methyl, CH_3I	Halogenated (CH_3I)
C–H methyl C–H, iodine (CH_3I)	1694	5902	C–H methyl, CH_3Br	Halogenated (CH_3I)
C–H methyl C–H, iodine (CH_3I)	1711	5845	C–H methyl, CH_3Br	Halogenated (CH_3I)

Functional Group	Nanometers (nm)	Wavenumbers (cm⁻¹)	Spectra-Structure	Material Type
C–H methyl C–H, nitro (CH$_3$NO$_2$)	1654	6045	C–H methyl, CH$_3$NO$_2$	Nitro (CH$_3$) as (CH$_3$NO$_2$)
C–H methyl C–H, nitro (CH$_3$NO$_2$)	1702	5874	C–H methyl, CH$_3$NO$_2$	Nitro (CH$_3$) as (CH$_3$NO$_2$)
C–H methyl C–H, OH associated as (ROHCH$_3$)	1664	6011	C–H methyl, OH associated as ROHCH$_3$	Alcohols, diols
C–H methyl C–H, OH associated as (ROHCH$_3$)	1688	5925	C–H methyl, OH associated as ROHCH$_3$	Alcohols, diols
C–H methyl C–H, OH associated as (ROHCH$_3$)	1701	5880	C–H methyl, OH associated as ROHCH$_3$	Alcohols, diols
C–H methyl C–H, OH associated as (ROHCH$_3$)	1732	5773	C–H methyl, OH associated as ROHCH$_3$	Alcohols, diols
C–H methylene (.CH$_2$)	762	13,123	C–H (5v), methylene C–H	Hydrocarbons, aliphatic
C–H methylene (.CH$_2$)	930	10,811–10,695	C–H (4v), .CH$_2$	Hydrocarbons, methylene
C–H methylene (.CH$_2$)	930	10,753	C–H (4v), methylene C–H	Hydrocarbons, aliphatic
C–H methylene (.CH$_2$)	1211	8258	C–H (3v), methylene C–H	Hydrocarbons, aliphatic
C–H methylene (.CH$_2$)	1215	8230	C–H (3v), .C–H$_2$	Hydrocarbons, aliphatic
C–H methylene (.CH$_2$)	1395	7168	C–H combination, .C–H$_2$	Hydrocarbons, aliphatic
C–H methylene (.CH$_2$)	1415	7067	C–H combination, .C–H$_2$	Hydrocarbons, aliphatic
C–H methylene (.CH$_2$)	1440	6944	C–H combination, .C–H$_2$	Hydrocarbons, aliphatic
C–H methylene (.CH$_2$)	1725	5797	C–H (2v), .CH$_2$	Hydrocarbons, methylene
C–H methylene (.CH$_2$)	1765	5666	C–H (2v), .CH$_2$	Hydrocarbons, methylene
C–H methylene (.CH$_2$)	1780	5618	C–H (2v), .CH$_2$	Cellulose
C–H methylene (.CH$_2$) (asymmetric)	1727	5787	C–H (2v), methylene C–H (asymmetric)	Hydrocarbons, aliphatic
C–H methylene (.CH$_2$) (symmetric)	1762	5675	C–H (2v), methylene C–H (symmetric)	Hydrocarbons, aliphatic
C–H methylene C–H, associated with branched aliphatic RC(CH$_3$)$_3$ or RCH(CH$_3$)$_2$	836	11,962	C–H (4vCH$_2$ and δCH$_2$) combination	Hydrocarbons, aliphatic
C–H methylene C–H, associated with branched aliphatic RC(CH$_3$)$_3$ or RCH(CH$_3$)$_2$	1042	9600	C–H (3vCH$_2$ and δCH$_2$) combination	Hydrocarbons, aliphatic

C–H methylene C–H, associated with branched aliphatic $RC(CH_3)_3$ or $RCH(CH_3)_2$	1411	7085	C–H ($2vCH_2$ and δCH_2) combination	Hydrocarbons, aliphatic
C–H methylene C–H, associated with branched aliphatic $RC(CH_3)_3$ or $RCH(CH_3)_2$	2318	4314	C–H ($2vCH_2$ asymmetric stretching and δCH_2) combination	Hydrocarbons, aliphatic
C–H methylene C–H, associated with branched aliphatic $RC(CH_3)_3$ or $RCH(CH_3)_2$	2363	4232	C–H ($2vCH_2$ symmetric stretching and δCH_2) combination	Hydrocarbons, aliphatic
C–H methylene C–H, associated with linear aliphatic $R(CH_2)_NR$	830	12,048	C–H ($4vCH_2$ and δCH_2) combination	Hydrocarbons, aliphatic
C–H methylene C–H, associated with linear aliphatic $R(CH_2)_NR$	1041	9606	C–H ($3vCH_2$ and δCH_2) combination	Hydrocarbons, aliphatic
C–H methylene C–H, associated with linear aliphatic $R(CH_2)_NR$	1410	7092	C–H ($2vCH_2$ and δCH_2) combination	Hydrocarbons, aliphatic
C–H methylene C–H, associated with linear aliphatic $R(CH_2)_NR$	2308	4333	C–H ($2vCH_2$ asymmetric stretching and δCH_2) combination	Hydrocarbons, aliphatic
C–H methylene C–H, associated with linear aliphatic $R(CH_2)_NR$	2347	4261	C–H ($2vCH_2$ symmetric stretching and δCH_2) combination	Hydrocarbons, aliphatic
C–H methylene C–H, associated with ovalbumin protein side chains seen at pH 5.0	2345	4265	C–H ($2vCH_2$ symmetric stretching and δCH_2) combination band from ovalbumin protein side chains seen at pH 5.0	C–H from ovalbumin protein side chains
C–H methyne C–H as (R–C–C≡C–H)	1530	6536	C–H, methyne (1-hexyne) as (R–C–C≡C–H)	Acetylene or Methyne C–H
C–H secondary or tertiary carbon (.CH)	1225	8163	C–H ($3v$), .C–H	Hydrocarbons, aliphatic
C–H starch (.C–H and CH_2)	2280	4386	C–H stretching and CH_2 deformation	Polysaccharides
C–H vinyl and vinylidene C–H as (CH_2=C(CH_3)–CH=CH_2)	1630	6135	C–H, vinyl and vinylidene (2-methyl-1,3-butadiene), isoprene as (CH_2=C(CH_3)–CH=CH_2)	Vinyl and vinylidene C–H as isoprene
C–H vinyl C–H attached to >C=O group, such as (CH_2=CH–C=O –)	1613	6200	C–H, vinyl on O (1-ethenyloxybutane) as (CH_2=CH–C=O –)	Vinyl as 1-ethenyl oxybutane (1-ethenyl-methyl-ethyl-ketone)

(continued)

Functional Group	Nanometers (nm)	Wavenumbers (cm⁻¹)	Spectra-Structure	Material Type
C–H vinyl C–H attached to N group, such as (CH₂=CH–N–C=O–C)	1621	6170	C–H, vinyl on N (1-ethenyl 2-pyrrolidinone) as (CH₂=CH–N–C=O–C)	Vinyl on pyrrolidinone
C–H vinyl C–H, associated with (CH₂=CH–)	1635	6120	C–H, vinyl (hexene), CH₂=CH–	Hydrocarbons, aliphatic
C–H vinylidene C–H, associated with (CH₂=C<)	1631	6130	C–H, vinylidene (secondary alkyl group), CH₂=C<	Hydrocarbons, aliphatic
C–H/C=O lipid associated (RC=CH and RC=O)	2140	4673	C–H stretching and C=O stretching combination and C–H deformation combination	Lipids
C–H/C–C (.C–H and .C–C)	2380	4202	C–H stretching and C–C stretching combination	Lipids
C–H/C–C (.C–H and .C–C)	2488	4019	C–H stretching and C–C stretching combination	Cellulose
C–H/C–C/C–O–C (.C–H and .C–C and .C–O–C)	2500	4000	C–H stretching and C–C and C–O–C stretching combination	Polysaccharides
CHO — classic filter instrument	2100	4762	Classic filter instrument	Carbohydrate
CHO — classic filter instrument	2230	4484	Classic filter instrument	Reference (classic)
CHO — classic filter instrument	2270	4405	Classic filter instrument	Lignin
CHO — classic filter instrument	2310	4329	Classic filter instrument	Lipids/oils
CHO — classic filter instrument	2336	4281	Classic filter instrument	Cellulose
CHO carbohydrate (.CHO)	2200	4545	C–H stretching and C=O combination	Carbohydrates
C–N–C amide: (.C–N–C)	2470	4049	C–N–C (2v)	Proteins
C–N–C amide: (.C–N–C)	2530	3953	C–N–C (2v) asymmetric	Amide
C=O (carbonyl >C=O)	1160	8621	C=O (5v)	Hydrocarbons, aliphatic
C=O (carbonyl >C=O)	1450	6897	C=O (4v)	Ketones and aldehydes
C=O amide (.C=ONH)	1920	5208	C=O (3v), .C=ONH	Amide
C=O amide: NH₂–C=ONH₂	2030	4926	C=O (3v), .C=ONH₂	Urea
C=O carbonyl (.C=OOH)	1900	5263	C=O (3v), .C=OOH	Acids, carboxylic
C=O esters and acids (.C=OOR)	1950	5128	C=O (3v), .C=OOR	Acids and esters
C=O/C–N/N–H combination from polyamide 11	2212	4521	C=O/C–N/N–H [2 × amide I (2vC=O stretching) and amide III (C–N stretching/N–H in-plane bending) combination] from polyamide 11	Polyamide 11
C=O/N–H combination from polyamide 11	1598	6256	C=O/N–H [amide I (2v$_S$C=O stretching) and 3 × amide II (N–H in-plane bending) combination] from polyamide 11	Polyamide 11

C=O–O polymeric (C=O and C–O stretching)	2100	C=O–O (4v)	4762	Polysaccharides
CONH$_2$ (HN...O=C band) of amide A	3030	CONH$_2$ as amide A for the polypeptides	3300	Amide A from proteins (polypeptides)
CONH$_2$ as combination of amide B and amide II modes (amide B/amide II) from ovalbumin	2174	CONH$_2$ as combination of amide B and amide II modes (amide B/amide II)	4600	Combination of amide B and amide II modes (amide B/amide II)
CONH$_2$ specifically due to amide B and amide II modes (amide B/II) from ovalbumin protein side chains seen at shifting to lower wavenumber with considerable broadening at lowered pH (from pH 5.0 to 2.4)	2174	CONH$_2$ specifically due to amide B and amide II modes (amide B/II)	4600	Proteins at low pH
CONH$_2$ specifically due to C=O hydrogen bonded to the N–H of the peptide link termed the α-helix structure	1738	CONH$_2$ specifically due to the α-helix peptide structure	5755	Proteins as normalized 2nd-derivative spectra of proteins in aqueous solution
CONH$_2$ specifically due to C=O hydrogen bonded to the N–H of the peptide link termed the α-helix structure	2167	CONH$_2$ specifically due to the α-helix peptide structure	4615	Proteins as normalized 2nd-derivative spectra of proteins in aqueous solution
CONH$_2$ specifically due to C=O hydrogen bonded to the N–H of the peptide link termed the α-helix structure	2290	CONH$_2$ specifically due to the α-helix peptide structure	4365–4370	Proteins as normalized 2nd-derivative spectra of proteins in aqueous solution
CONH$_2$ specifically due to C=O hydrogen bonded to the N–H of the peptide link termed the α-helix structure	2445	CONH$_2$ specifically due to the α-helix peptide structure	4090	Proteins as normalized 2nd-derivative spectra of proteins in aqueous solution
CONH$_2$ specifically due to peptide N–H and C=O groups at right angles to the line of the peptide backbone referred to as the β-sheet structure	1690	CONH$_2$ specifically due to peptide β-sheet structures	5915–5925	Proteins as normalized 2nd-derivative spectra of proteins in aqueous solution

(continued)

Functional Group	Nanometers (nm)	Wavenumbers (cm⁻¹)	Spectra-Structure	Material Type
CONH₂ specifically due to peptide N–H and C=O groups at right angles to the line of the peptide backbone referred to as the β-sheet structure	2055	4865	CONH₂ specifically due to peptide β-sheet structures	Proteins as normalized 2nd-derivative spectra of proteins in aqueous solution
CONH₂ specifically due to peptide N–H and C=O groups at right angles to the line of the peptide backbone referred to as the β-sheet structure	2208	4525–4540	CONH₂ specifically due to peptide β-sheet structures	Proteins as normalized 2nd-derivative spectra of proteins in aqueous solution
CONH₂ specifically due to peptide N–H and C=O groups at right angles to the line of the peptide backbone referred to as the β-sheet structure	2270	4405	CONH₂ specifically due to peptide β-sheet structures	Proteins as normalized 2nd-derivative spectra of proteins in aqueous solution
CONH₂ specifically due to peptide N–H and C=O groups at right angles to the line of the peptide backbone referred to as the β-sheet structure	2463	4060	CONH₂ specifically due to peptide β-sheet structures	Proteins as normalized 2nd-derivative spectra of proteins in aqueous solution
HN...O=C band of amide A	2060	4855	CONH₂ combination of amide A and amide II	Combination of amide A and amide II from proteins (polypeptides)
N–H — classic filter instrument	2180	4587	N–H classic filter instrument	Protein
N–H from primary and secondary aromatic amine (mixtures of aniline and N-ethylaniline in CCl₄)	1493	6698	N–H (2ν) symmetric, primary aromatic amine in CCl₄ as ortho-OCH3 substituent	Aromatic amines
N–H primary aromatic amine (m-Cl)	1018.5	9818	N–H (3ν) asymmetric, primary aromatic amine in CCl₄ as meta-Cl substituent	Aromatic amine
N–H primary aromatic amine (m-Cl)	1445.0	6920	N–H (2ν) asymmetric, primary aromatic amine in CCl₄ as meta-Cl substituent	Aromatic amine
N–H primary aromatic amine (m-Cl)	1489.5	6713	N–H (2ν) symmetric, primary aromatic amine in CCl₄ as meta-Cl substituent	Aromatic amine

(continued)

N–H primary aromatic amine (m-NO₂)	1015.0	9852	N–H (3v) asymmetric, primary aromatic amine in CCl₄ as *meta*-NO₂ substituent	Aromatic amine
N–H primary aromatic amine (m-NO₂)	1441.0	6940	N–H (2v) asymmetric, primary aromatic amine in CCl₄ as *meta*-NO₂ substituent	Aromatic amine
N–H primary aromatic amine (m-NO₂)	1487.0	6725	N–H (2v) symmetric, primary aromatic amine in CCl₄ as *meta*-NO₂ substituent	Aromatic amine
N–H primary aromatic amine (no other substituents)	1021.0	9794	N–H (3v) asymmetric, primary aromatic amine in CCl₄, no other substituents	Aromatic amine
N–H primary aromatic amine (no other substituents)	1448.5	6904	N–H (2v) asymmetric, primary aromatic amine in CCl₄, no other substituents	Aromatic amine
N–H primary aromatic amine (no other substituents)	1493.0	6698	N–H (2v) symmetric, primary aromatic amine in CCl₄, no other substituents	Aromatic amine
N–H primary aromatic amine (o-Cl)	1017.5	9828	N–H (3v) asymmetric, primary aromatic amine in CCl₄ as *ortho*-Cl grouping	Aromatic amine
N–H primary aromatic amine (o-Cl)	1443.0	6930	N–H (2v) asymmetric, primary aromatic amine in CCl₄ as *ortho*-Cl grouping	Aromatic amine
N–H primary aromatic amine (o-Cl)	1491.5	6705	N–H (2v) symmetric, primary aromatic amine in CCl₄ as *ortho*-Cl grouping	Aromatic amine
N–H primary aromatic amine (o-NO₂)	1003.0	9970	N–H (3v) asymmetric, primary aromatic amine in CCl₄ as *ortho*-NO₂ substituent	Aromatic amine
N–H primary aromatic amine (o-NO₂)	1432.0	6982	N–H (2v) asymmetric, primary aromatic amine in CCl₄ as *ortho*-NO₂ substituent	Aromatic amine
N–H primary aromatic amine (o-NO₂)	1472	6791	N–H (2v) symmetric, primary aromatic amine in CCl₄ as *ortho*-NO₂ substituent	Aromatic amine
N–H primary aromatic amine (o-OCH₃)	1018.5	9818	N–H (3v) asymmetric, primary aromatic amine in CCl₄ as *ortho*-OCH3 substituent	Aromatic amine
N–H primary aromatic amine (o-OCH₃)	1446.0	6916	N–H (2v) asymmetric, primary aromatic amine in CCl₄ as *ortho*-OCH3 substituent	Aromatic amine
N–H primary aromatic amine (o-OCH₃)	1492.5	6700	N–H (2v) symmetric, primary aromatic amine in CCl₄ as *ortho*-OCH3 substituent	Aromatic amine
N–H primary aromatic amine (p-CH₃)	1023.0	9775	N–H (3v) asymmetric, primary aromatic amine in CCl₄ as *para*-CH₃ grouping	Aromatic amine
N–H primary aromatic amine (p-CH₃)	1452.5	6885	N–H (2v) asymmetric, primary aromatic amine in CCl₄ as *para*-CH₃ grouping	Aromatic amine

Functional Group	Nanometers (nm)	Wavenumbers (cm^{-1})	Spectra-Structure	Material Type
N–H primary aromatic amine (p-CH$_3$)	1496.5	6683	N–H (2ν) symmetric, primary aromatic amine in CCl$_4$ as $para$-CH$_3$ grouping	Aromatic amine
N–H primary aromatic amine (p-Cl)	1019.0	9813	N–H (3ν) asymmetric, primary aromatic amine in CCl$_4$ as $para$-Cl substituent	Aromatic amine
N–H primary aromatic amine (p-Cl)	1448.0	6906	N–H (2ν) asymmetric, primary aromatic amine in CCl$_4$ as $para$-Cl substituent	Aromatic amine
N–H primary aromatic amine (p-Cl)	1491.5	6705	N–H (2ν) symmetric, primary aromatic amine in CCl$_4$ as $para$-Cl substituent	Aromatic amine
N–H primary aromatic amine (p-NH$_2$)	1026.5	9741	N–H (3ν) asymmetric, primary aromatic amine in CCl$_4$ as $para$-NH$_2$ grouping	Aromatic amine
N–H primary aromatic amine (p-NH$_2$)	1459.5	6852	N–H (2ν) asymmetric, primary aromatic amine in CCl$_4$ as $para$-NH$_2$ grouping	Aromatic amine
N–H primary aromatic amine (p-NH$_2$)	1502.5	6656	N–H (2ν) symmetric, primary aromatic amine in CCl$_4$ as $para$-NH$_2$ grouping	Aromatic amine
N–H amide .NH	1570	6369	N–H (2ν), .CONHR	Amide/protein
N–H amide .NH or .NH$_2$	1460	6849	N–H (2ν), symmetrical	Urea
N–H amide .NH or .NH$_2$	1463	6835	N–H (2ν), .CONH$_2$	Amide/protein
N–H amide .NH or .NH$_2$	1483	6743	N–H (2ν), .CONH$_2$	Amide/protein
N–H amide .NH or .NH$_2$	1492	6702	N–H (2ν), ArNH$_2$	Amide/protein
N–H amide .NH or .NH$_2$	1500	6667	N–H (2ν), .NH$_2$	Amide/protein
N–H amide .NH or .NH$_2$	1510	6623	N–H (2ν), .CONH$_2$	Amide/protein
N–H amide .NH or .NH$_2$	1520	6579	N–H (2ν), .CONH$_2$	Amide/protein
N–H amide .NH or .NH$_2$	1530	6536	N–H (2ν), RNH$_2$	Amide/protein
N–H amide as (.CONH.) and (.CONH$_2$)	2060	4854	N–H (3δ) and N–H stretching combination	Amides/proteins
N–H amide II (.CONH$_2$)	1980	5051	N–H stretching (asymmetric) and N–H in-plane bending combination	Amides/proteins
N–H amide with N–R group	1471	6798	N–H (2ν), .CONHR	Amide/protein
N–H amide with N–R group	1490	6711	N–H (2ν), .CONHR	Amide/protein
N–H amide: (N–H deformation)	2070	4831	N–H (δ)	Urea
N–H amide: NH$_2$–C=ONH$_2$	1990	5025	N–H stretching and N–H bending combination	Urea
N–H ammonia in water	1534	6250	N–H (2ν) for NH$_3$ (ammonia) in water	Ammonia in water

N–H ammonia in water	2000	5000	N–H (vN–H and δN–H combination) for NH_3 (ammonia) in water	N–H (combination band) for ammonia in water
N–H ammonia in water	2210	4525	N–H (3v) for NH_3 (ammonia) in water	Ammonia in water
N–H band found at 4867 cm^{-1} for native RNase A shifting to 4878 cm^{-1} upon thermal unfolding	2053	4867–4878	N–H from $CONH_2$ as thermal unfolding of RNase A protein in aqueous solution (assigned to an N–H combination band)	Protein thermal unfolding
N–H band from urea (NH_2–C=O–NH_2)	1470	6803	N–H (2v) asymmetric stretching from urea	N–H from urea
N–H band from urea (NH_2–C=O–NH_2)	1485	6736	N–H (2v) symmetric stretching from urea	N–H from urea
N–H bonded combination from polyamide 11	1570	6368	N–H [bonded NH stretching and 2 × amide II deformation (N–H in-plane bending) combination] from polyamide 11	Polyamide 11
N–H bonded from polyamide 11	1515	6600	N–H (2v) stretching bonded NH, disordered phase from polyamide 11	Polyamide 11
N–H bonded from polyamide 11	1538	6500	N–H (2v) stretching bonded NH, ordered phase from polyamide 11	Polyamide 11
N–H combination band (RNase A) – C=O amide I band	2050	4878	N–H native RNase A combination band at 4867 cm^{-1} shifting due to thermal unfolding (C=O amide I band)	N–H from protein
N–H combination band (RNase A) – C=O amide I band	2055	4867	N–H combination band found in the spectrum of native RNase A (C=O amide I band)	N–H from protein
N–H combination band from primary amides (R–C=O–NH_2)	1470	6805	N–H (vN–H asymmetric and vN–H symmetric combination) for primary amides	N–H combination band from primary amides
N–H combination band from primary amides (R–C=O–NH_2)	1960	5100	N–H [vN–H asymmetric and amide II deformation (N–H in-plane bending) combination] for primary amides	N–H combination band from primary amides
N–H combination band from secondary amides in native RNase A	2075	4820	N–H [vN–H and amide II deformation (N–H in-plane bending) combination] for secondary amides in native RNase A	N–H from native RNase A
N–H combination band from secondary amides in proteins	2060	4850	N–H [vN–H and amide II deformation (N–H in-plane bending) combination] for secondary amides in proteins	N–H from protein
N–H combination band from urea (NH_2–C=O–NH_2)	1500	6666	N–H (vN–H asymmetric and vN–H symmetric combination) from urea	N–H from urea

(continued)

Functional Group	Nanometers (nm)	Wavenumbers (cm⁻¹)	Spectra-Structure	Material Type
N–H combination band from urea (NH_2–C=O–NH_2)	2040	4902	N–H [νN–H symmetric and amide II deformation (N–H in-plane bending) combination] for primary amides	N-H from urea
N–H combination band from urea (NH_2–C=O–NH_2)	2220	4505	N–H (νN–H asymmetric and NH_2 rocking) combination	N-H from urea
N–H combination from polyamide 11	2053	4870	Bonded NH stretching and amide II (N–H in-plane bending) from polyamide 11	Polyamide 11
N–H combination, from primary and secondary aromatic amine (mixtures of aniline and N-ethylaniline in CCl_4)	1972	5071	N–H (νN–H and δN–H combination), primary aromatic amine in CCl_4 as para-Cl substituent from primary and secondary aromatic amines	Primary and secondary aromatic amines
N–H combination, primary aromatic amine (m-Cl)	1968.0	5082	N–H (νN–H and δN–H combination), primary aromatic amine in CCl_4 as meta-Cl substituent	Aromatic amine
N–H combination, primary aromatic amine (m-NO_2)	1963.0	5095	N–H (νN–H and δN–H combination), primary aromatic amine in CCl_4 as meta-NO_2 substituent	Aromatic amine
N–H combination, primary aromatic amine (no other substituents)	1971.5	5072	N–H (νN–H and δN–H combination), primary aromatic amine in CCl_4, no other substituents	Aromatic amine
N–H combination, primary aromatic amine (o-Cl)	1971.5	5072	N–H (νN–H and δN–H combination), primary aromatic amine in CCl_4 as ortho-Cl grouping	Aromatic amine
N–H combination, primary aromatic amine (o-NO_2)	1967.0	5084	N–H (νN–H and δN–H combination), primary aromatic amine in CCl_4 as ortho-NO_2 substituent	Aromatic amine
N–H combination, primary aromatic amine (o-OCH_3)	1977.5	5057	N–H (νN–H and δN–H combination), primary aromatic amine in CCl_4 as ortho-OCH3 substituent	Aromatic amine
N–H combination, primary aromatic amine (p-CH_3)	1975.5	5062	N–H (νN–H and δN–H combination), primary aromatic amine in CCl_4 as para-CH_3 grouping	Aromatic amine
N–H combination, primary aromatic amine (p-Cl)	1970.5	5075	N–H (νN–H and δN–H combination), primary aromatic amine in CCl_4 as para-Cl substituent	Aromatic amine
N–H combination, primary aromatic amine (p-NH_2)	1980.5	5049	N–H (νN–H and δN–H combination), primary aromatic amine in CCl_4 as para-NH_2 grouping	Aromatic amine
N–H for secondary amine as (R–NH–R)	1461	6844	N–H (2ν), secondary amine as (R–NH–R), indole, first overtone band intensity comparisons, CCl_4 solution	N-H secondary amine

(continued)

N–H for secondary amine as (R–NH–R)	1465	6826	N–H (2v), secondary amine as (R–NH–R), carbazole, first overtone band intensity comparisons, CCl$_4$ solution	N–H secondary amine
N–H for secondary amine as (R–NH–R)	1481	6751	N–H (2v), secondary amine as (R–NH–R), N-benzylamine, first overtone band intensity comparisons, CCl$_4$ solution	N–H secondary amine
N–H for secondary amine as (R–NH–R)	1486	6729	N–H (2v), secondary amine as (R–NH–R), diphenylamine, first overtone band intensity comparisons, CCl$_4$ solution	N–H secondary amine
N–H for secondary amine as (R–NH–R)	1520	6580	N–H (2v), secondary amine as (R–NH–R), dimethylamine (vapor), first overtone band intensity comparisons, CCl$_4$ solution	N–H secondary amine
N–H for secondary amine as (R–NH–R)	1530	6536	N–H (2v), secondary amine as (R–NH–R), morpholine, first overtone band intensity comparisons, CCl$_4$ solution	N–H secondary amine
N–H for secondary amine as (R–NH–R)	1545	6471	N–H (2v), secondary amine as (R–NH–R), diethylamine, first overtone band intensity comparisons, CCl$_4$ solution	N–H secondary amine
N–H from gamma-valerolactam	2055	4865	N–H (vN–H and δN–H combination) for gamma-valerolactam	Gamma-valerolactam
N–H from gamma-valerolactam	2090	4785	N–H [NH stretching and amide II deformation (N–H in-plane bending) combination] for gamma-valerolactam	Gamma-valerolactam
N–H from polyamide 11	1618	6180	N–H {4 × amide II (N–H in-plane bending)] from polyamide 11	Polyamide 11
N–H from primary amides (R–C=O–NH$_2$)	1430	6995	N–H (2v) symmetric from primary amide (R–C=O–NH$_2$)	N–H from primary amides
N–H from primary amides (R–C=O–NH$_2$)	1490	6710	N–H (2v) asymmetric from primary amide (R–C=O–NH$_2$)	N–H from primary amides
N–H from secondary amide in proteins	1529–1600	6540–6250	N–H (2v) stretching from secondary amide in proteins	N–H from protein
N–H nonbonded from polyamide 11	1480	6760	N–H (2v) stretching nonbonded N–H from polyamide 11	Polyamide 11

Functional Group	Nanometers (nm)	Wavenumbers (cm^{-1})	Spectra-Structure	Material Type
N–H primary amine band of diamino- compound	2205	4535	N–H (3ν) band of primary amine as bisphenol A (I) resins cured with 4,4'-diaminodiphenyl sulfone (II) hardener	Amine, primary
N–H proteins: N–H ($3\nu_B$)	2180	4587	N–H (3δ)	Proteins/amino acids
N–H/C=O amide as (.CONH .) and (.CONH$_2$)	2055	4866	N–H stretching and C=O stretching (amide I) combination	Amides/proteins
N–H/C=O amide as (.CONH .) and (.CONH$_2$) from native RNase A	2055	4867	N–H stretching and C=O stretching (amide I) combination band in the spectrum of native RNase A	RNase A
N–H/C=O amide as (.CONH .) and (.CONH$_2$) from native RNase A (thermal unfolding)	2050	4878	N–H stretching and C=O stretching (amide I) combination band observed in the thermal unfolding observed in native RNase A	RNase A
N–H/C=O combination from native RNase A	2024	4940	N–H/C=O [bonded NH stretching and amide I (2νC=O stretching) combination] of native RNase A	RNase A
N–H/C=O combination from polyamide 11	2012	4970	N–H/C=O [bonded NH stretching and amide I (2νC=O stretching) combination] from polyamide 11	Polyamide 11
N–H/C=O combination from polyamide 11	2127	4701	N–H/C=O [2 × amide II (N–H in-plane bending) and amide I (2νC=O stretching) combination] from polyamide 11	Polyamide 11
N–H/C–N combination band from primary amides (R–C=O–NH$_2$)	2010	4975	N–H [νN–H symmetric and amide III deformation (C–N stretching/N–H in-plane bending) combination] for primary amides	N–H combination band from primary amides
N–H/C–N combination band from primary amides (R–C=O–NH$_2$)	2030	4925	N–H [νN–H asymmetric and amide III deformation (C–N stretching/N–H in-plane bending) combination] for primary amides	N–H combination band from primary amides
N–H/C–N combination band from urea (NH$_2$–C=O–NH$_2$)	1990	5025	N–H [νN–H asymmetric and amide III deformation (C–N stretching/N–H in-plane bending) combination] for primary amides	N–H from urea

N–H/C–N combination band from urea (NH_2–C=O–NH_2)	2080	N–H/C–N [vN–H asymmetric and amide III deformation (C–N stretching/N–H in-plane bending) combination] for urea	4808	N–H/C–N from urea
N–H/C–N from gamma-valerolactam	2120	N–H/C–N [vN–H and amide III deformation (C–N stretching/N–H in-plane bending) combination] for gamma-valerolactam	4715	Gamma-valerolactam
N–H/C–N/C=O combination band from secondary amides in proteins	2180	N–H/C–N/C=O [2 × amide I (2vC=O stretching) and amide III deformation (C–N stretching/N–H in-plane bending) combination] for secondary amides in proteins	4590	N–H/C–N/C=O from protein
N–H/C–N/C=O combination band from urea (NH_2–C=O–NH_2)	2180	N–H/C–N/C=O [2 × amide I (2vC=O stretching) and amide III deformation (C–N stretching/N–H in-plane bending) combination] for urea	4687	N–H/C–N/C=O from urea
N–H/C–N/C=O from gamma-valerolactam	2145	N–H/C–N/C=O [2 × amide I (2vC=O stretching) and amide III deformation (C–N stretching/N–H in-plane bending) combination] for gamma-valerolactam	4660	Gamma-valerolactam
N–H/C–N/N–H amide II and amide III combination (.CONH .) and ($.CONH_2$)	2050	N–H in-plane bend and C–N stretching and N–H in-plane bend combination	4878	Amides/proteins
N–H/C–N/N–H combination from polyamide 11	2183	N–H/C–N/N–H [bonded NH stretching and amide III (C–N stretching/N–H in-plane bending) combination] from polyamide 11	4586	Polyamide 11
OH — classic filter instrument	1940	Classic filter instrument	5155	Moisture (classic)
O–H and CH combination from methanol	1965	v O–H and δC–H combination from CH_3OH	5090	Methanol
O–H and C–O stretching and bending combination from methanol	2080	v O–H and δO–H combination from CH_3OH	4808	Methanol
O–H (.O–H and HOH)	1927–1929	O–H stretching and HOH deformation combination from water molecules in the 3-aminopropyltriethoxysilane-ethanol-water system	5189–5184	O–H stretching and HOH deformation combination from 3-aminopropyltriethoxysilane-ethanol-water system
O–H (.O–H and HOH)	1930	O–H stretching and HOH bending combination	5181	Polysaccharides
O–H (H–O–H) Water	1940	O–H stretching and HOH bending combination	5155	Water
O–H alcohol (RO–H)	1410	O–H (2v), .O–H	7092	Hydrocarbons, aliphatic

(continued)

Functional Group	Nanometers (nm)	Wavenumbers (cm⁻¹)	Spectra-Structure	Material Type
O–H alcohol O–H with hydrogen bonding	(1416)	(7060 primarily)	O–H (2v) from butanol	Alkyl alcohols (containing one O–H)
O–H alcohol O–H with hydrogen bonding	1389–1426	7200–7015	O–H (2v) from hydrogen-bonded alkyl alcohols as (R–C–OH)	Alkyl alcohols (containing one O–H)
O–H alcohol O–H with hydrogen bonding	1460–1603	6850–6240		
O–H alcoholic O–H, intramolecularly bonded	1460	6850	O–H (2v) intramolecularly bonded from diols	Alkyl alcohols (diols)
O–H alcoholic O–H, nonhydrogen bonded	1415	7065	O–H (2v) nonbonded from diols	Alkyl alcohols (diols)
O–H alkyl alcohols O–H with no hydrogen bonding (R–C–OH) in CCl₄	513	19,500	O–H (6v) from nonhydrogen-bonded short-chain alkyl alcohols in CCl₄ as (R–C–OH)	Alkyl alcohols
O–H alkyl alcohols O–H with no hydrogen bonding (R–C–OH) in CCl₄	599	16,700	O–H (5v) from nonhydrogen-bonded short-chain alkyl alcohols in CCl₄ as (R–C–OH)	Alkyl alcohols
O–H alkyl alcohols O–H with no hydrogen bonding (R–C–OH) in CCl₄	741	13,500	O–H (4v) from nonhydrogen-bonded short-chain alkyl alcohols in CCl₄ as (R–C–OH)	Alkyl alcohols
O–H alkyl alcohols O–H with no hydrogen bonding (R–C–OH) in CCl₄	962	10,400	O–H (3v) from nonhydrogen-bonded short-chain alkyl alcohols in CCl₄ as (R–C–OH)	Alkyl alcohols
O–H aromatic (ArO–H)	1420	7042	O–H (2v), .O–H	Hydrocarbons, aromatic
O–H assigned to molecular water [O–H (.O–H and HOH)]	1923	5200	O–H assigned to molecular water (O–H stretching and HOH deformation combination)	O–H molecular water
O–H-bound water and the first overtone of an NH-stretching mode of the amide groups of ovalbumin	1550	6450	O–H-bound water with several hydrogen bonds and the first overtone of an NH-stretching mode of the amide groups of ovalbumin	O–H-bound water of the amide groups of ovalbumin
O–H broad band occurring in polyols, alcohols, water, ethylene vinyl alcohols and copolymers containing O–H groups	1802–2198	5550–4550 (broad)	O–H broad band, due to vO–H and δO–H combination	Polyols, alcohols, water

O–H combination band, alcohols or water	1065	9386	2vO–H and δC-H methyl combination	Alcohols as R–C–O–H
O–H combination band, alcohols or water	1580	6330	2vO–H, O–H broad band, hydrogen bonded	Alcohols as R–C–O–H
O–H deformation band, alcohols or water	2096	4770	3δO–H of mid-infrared band occurring near 1420 cm⁻¹	Alcohols or water O–H
O–H stretching band, alkyl alcohols or water	1583	6319	vO–H and δO–H combination reported in the literature from ethylene-vinyl alcohol copolymer spectra, probably better assigned as 2vO–H	Alcohols or water O–H
O–H stretching band, alcohols or water	1408 and 1603	7100 and 6240	2vO–H of different aggregates, such as monomers, dimers, and polyols	Alcohols or water O–H
O–H from alcohols in CCl$_4$ (first overtone (2v$_3$) of the O–H alcoholic stretching modes	1404–1422	7120–7030	O–H from alcohols in CCl$_4$ as n-, sec-, and $tert$-butanol. The three kinds of alcohols showed an intense band due to the first overtone (2v) of the OH stretching modes.	Alcohols in CCl$_4$
O–H from primary alcohols as (–CH$_2$–OH)	767	13,038	O–H (4v) (–CH$_2$–OH), primary alcohols	Primary alcohols
O–H from primary alcohols as (–CH$_2$–OH)	996	10,040	O–H (3v) (–CH$_2$–OH), primary alcohols	Primary alcohols
O–H from primary alcohols as (–CH$_2$–OH)	1452	6887	O–H (2v) (–CH$_2$–OH), primary alcohols	Primary alcohols
O–H from primary alcohols as (–CH$_2$–OH)	2740	3650	O–H (v) (–CH$_2$–OH), primary alcohols	Primary alcohols
O–H from secondary alcohols as (–CH–OH)	773	12,937	O–H (4v) (–CH–OH), secondary alcohols	Secondary alcohols
O–H from secondary alcohols as (–CH–OH)	1004	9960	O–H (3v) (–CH–OH), secondary alcohols	Secondary alcohols
O–H from secondary alcohols as (–CH–OH)	1464	6831	O–H (2v) (–CH–OH), secondary alcohols	Secondary alcohols
O–H from secondary alcohols as (–CH–OH)	2762	3620	O–H (v) (–CH–OH), secondary alcohols	Secondary alcohols
O–H from sugar as crystalline sucrose	1441	6940	O–H (2v) carbohydrates (C4 hydroxyl within a crystalline matrix)	Crystalline sucrose

(continued)

Functional Group	Nanometers (nm)	Wavenumbers (cm⁻¹)	Spectra-Structure	Material Type
O–H from tertiary alcohols as (–C–OH)	776	12,887	O–H (4v) (–C–OH), tertiary alcohols	Tertiary alcohols
O–H from tertiary alcohols as (–C–OH)	1006	9940	O–H (3v) (–C–OH), tertiary alcohols	Tertiary alcohols
O–H from tertiary alcohols as (–C–OH)	1468	6812	O–H (2v) (–C–OH), tertiary alcohols	Tertiary alcohols
O–H from tertiary alcohols as (–C–OH)	2770	3610	O–H (v) (–C–OH), tertiary alcohols	Tertiary alcohols
O–H from water	1790	5587	O–H combination	Water
O–H hydrogen bonding between water and exposed polyvinyl alcohol OH	1892	5285	O–H assigned as a one-to-one hydrogen-bonded to an isolated OH in the manner as OH::OH$_2$ from the effect of the hydration of the isolated alcohol OH — this pertains to the interactions of water with the OH groups in poly(ethylene-co-vinyl alcohol) (EVOH)	Water and polyvinyl alcohol OH
O–H hydrogen bonding between water and exposed polyvinyl alcohol OH	1942	5150	O–H interaction of multiple hydrogen atoms from poly(ethylene-co-vinyl alcohol) (EVOH) bonded to surrounding associated OH groups (from water) without clustering	Water and polyvinyl alcohol OH
O–H hydroxyl O–H free form (not hydrogen bonded)	1410	7090	O–H, free hydroxyl group in dilute CCl$_4$	O–H not hydrogen bonded
O–H methanol O–H with hydrogen bonding	1408	7100	O–H (2v) from nonhydrogen-bonded methanol in CCl$_4$ as (CH$_3$–OH)	O–H from methanol (nonhydrogen bonded)
O–H methanol O–H with hydrogen bonding	1470	6800	O–H (2v) from nonhydrogen-bonded methanol in CCl$_4$ as (CH$_3$–OH)	O–H from methanol (nonhydrogen bonded)
O–H phenolic O–H	1401–1441	7140–6940	2vO–H from phenols and aryl alcohols Ar–OH	Phenolic O–H
O–H phenols, dilute phenol in CCl$_4$	746 and 782	13,400 and 12,788 (doublet)	4vO–H, *ortho*-iodophenol	Phenolic O–H
O–H phenols, dilute phenol in CCl$_4$	755	13,250	4vO–H dilute phenol in CCl$_4$	Phenolic O–H
O–H phenols, dilute phenol in CCl$_4$	782 and 746	12,788 and 13,400 (doublet)	4vO–H, *ortho*-iodophenol	Phenolic O–H

O–H phenols, dilute phenol in CCl₄	1000	10,000	3νO–H dilute phenol in CCl₄	Phenolic O–H
O–H phenols, dilute phenol in CCl₄	1009 and 972	9911 and 10,288 (doublet)	3νO–H, *ortho*-iodophenol	Phenolic O–H
O–H phenols, *ortho*-substituted halogens	1464	6830	2νO–H in *ortho*-substituted halogens on phenols	Phenolic O–H
O–H polymeric (.O–H)	1450	6897	O–H (2ν), .O–H	Starch/polymeric alcohol
O–H polymeric (.O–H)	1490	6711	O–H (2ν), .O–H	Starch/polymeric alcohol
O–H polymeric (.O–H)	1540	6494	O–H (2ν), .O–H	Starch/polymeric alcohol
O–H polymeric (.O–H)	1960	5102	O–H stretching and HOH bending combination	Polysaccharides
O–H polymeric (.O–H)	2090	4785	O–H combination	Polymeric .OH
O–H related combination from water change in phase and N–H/C–N combination band from urea (NH₂–C=O–NH₂) from ovalbumin	2083	4800	O–H related combination from water change in phase with the increase in protein concentration at pH above 2.8 overlapping with a band representing the N–H/C–N [νₛN–H asymmetric and amide III deformation (C–N stretching/N–H in-plane bending) combination]	O–H related combination from water change in phase and N–H/C–N from ovalbumin protein
O–H stretching and bending combination from methanol	2016	4960	ν O–H and δO–H combination from CH₃OH	Methanol O–H
O–H stretching vibration (ν) of water molecules in bound-water hydrogen bonding in turquoise minerals	3049	3280	O–H ν (as intense and very broad band assigned to bound-water hydrogen bonding in turquoise minerals from Arizona and Senegal with a formula of Cu(Al6-x,Fex)(PO4)4(OH)8.4H2O	Bound-water hydrogen bonding in minerals
O–H stretching vibration (ν) of water molecules in protein-water hydrogen bonding	3067	3260	O–Hν (as intense and very broad band assigned to protein-water hydrogen bonding	Protein-water hydrogen bonding
O–H stretching vibrations of water molecules in water-water hydrogen bonding	3521	2840	O–H as intense and very broad band assigned to water-water hydrogen bonding	Water-water hydrogen bonding
O–H with hydrogen bonding (R–C–OH)	1047	9550	O–H (3ν) from hydrogen-bonded short-chain alkyl compounds	Alkyl alcohols
O–H, monomeric phenol in CCl₄	1420	7040	2νO–H monomeric phenol in CCl₄	Phenolic O–H
O–H/C–H cellulose (.OH and .C–O)	2270	4405	O–H stretching and C–O stretching combination	Cellulose
O–H/C–H combination	1820	5495	O–H stretching and C–O stretching (3νₛ) combination	Cellulose

(continued)

Functional Group	Nanometers (nm)	Wavenumbers (cm^{-1})	Spectra-Structure	Material Type
O–H/C–O from glucose	2273	4400	O–H/C–O glucose absorption from O–H stretching and C–O stretching combination	Glucose
O–H/C–O polyfunctional alkyl alcohols	1029	9720	2vO–H and 3vC–O combination	Ethers and esters also containing alcohols
O–H/C–O polymeric (.O–H and .C–O)	2100	4762	O–H bending and C–O stretching combination	Polysaccharides
O–H/C–O stretching combination from methanol	2123	4710	vO–H and vC–O combination from CH$_3$OH	Methanol O–H
P–OH phosphate (.P–OH)	1908	5241	O–H (2v), .P–OH	Phosphate
Reference — classic filter instrument	1680	5952	CH – aromatic used as classic reference filter for filter instruments	Reference (classic)
S–H thiol (.S–H)	1740	5747	S–H (2v), .S–H	Thiols
Si–O from silicone	1452	6887	Si–O (2v) stretch from silicone (dimethyl siloxane)	Si–O from silicone (dimethyl siloxane)
Si–O–H stretch + Si–O–Si combination from silicone	1933	5173	Si–O–H stretch + Si–O–Si deformation combination from silicone (dimethyl siloxane)	Si–O–H stretch + Si–O–Si from silicone (dimethyl siloxane)

ABBREVIATIONS AND SYMBOLS

(1ν)	*Fundamental* stretching vibrational absorption band
(2ν)	*First* overtone of fundamental stretching band
(3ν)	*Second* overtone of fundamental stretching band
(4ν)	*Third* overtone of fundamental stretching band
(5ν)	*Fourth* overtone of fundamental stretching band
(6ν)	*Fifth* overtone of fundamental stretching band
(1δ)	*Fundamental* bending vibrational absorption band
(2δ)	*First* overtone of fundamental bending band
(3δ)	*Second* overtone of fundamental bending band
(4δ)	*Third* overtone of fundamental bending band
(ω)	Deformation (rocking or wagging)

Appendix 5: Spectra Index by Functional Group or Comparison Series Group

Functional Group	Compound Name	Molecular Formula	CAS Number	10,500 – 6300 cm^{-1} (952 – 1587 nm)	7200-3800 cm^{-1} (1389 – 2632 nm)	Other NIR Regions
Alcohol (mono-OH)	1-Butanol	$C_4H_{10}O$	71-36-3	119, 208, 209	167, 210	—
Alcohol (mono-OH)	1-Propanol	C_3H_8O	71-23-8	119,209	—	—
Alcohol (mono-OH)	2-Butanol	$C_4H_{10}O$	78-92-2	209	—	—
Alcohol (mono-OH)	2-methyl-2-Propanol	$C_4H_{10}O$	75-65-0	118, 119, 209, 234, 235	166,167, 236, 237	—
Alcohol (mono-OH)	2-Propanol	C_3H_8O	67-63-0	119, 209	167	—
Alcohol (mono-OH)	3-Pentanol	$C_5H_{12}O$	584-02-01	119, 209	167	—
Alcohol (mono-OH)	Methanol	CH_4O	67-56-1	119, 208, 209	167, 210	—
Alcohol (diol)	1,2-Ethanediol	$C_2H_6O_2$	107-21-1	120	168	—
Alcohol (diol)	1,2 Propanediol	$C_3H_8O_2$	57-55-6	120	168	—
Alcohol (diol)	2,3-Butanediol	$C_4H_{10}O_2$	513-85-9	120, 210	168, 211	—
Alcohol (diol)	2,4-Pentanediol	$C_5H_{12}O_2$	625-69-4	120	168	—
Alcohol (diol)	2,5-Hexanediol	$C_6H_{14}O_2$	2935-44-6	120	168	—
Butanediol Series	1,2-Butanediol	$C_4H_{10}O_2$	584-03-2	121	169, 211	—
Butanediol Series	1,3-Butanediol	$C_4H_{10}O_2$	107-88-0	121	169, 211	—
Butanediol Series	1,4-Butanediol	$C_4H_{10}O_2$	110-63-4	121	169, 211	—
Butanediol Series	2,3-Butanediol	$C_4H_{10}O_2$	513-85-9	121, 208	169	—
Ether	1,1,2-trimethoxy-Ethane	$C_5H_{12}O_3$	24332-20-5	122, 212	170, 213	—
Ether	1,1-dimethoxy-Ethane	$C_4H_{10}O_2$	534-15-6	122	170	—
Ether	1,2-diethoxy-Ethane	$C_6H_{14}O_2$	629-14-1	122	170	—
Ether	2,2-dimethoxy-Propane	$C_5H_{12}O_2$	77-76-7	122, 212	170, 213	—
Ether	2-ethoxy-Ethanol	$C_4H_{10}O_2$	110-80-5	123	171	—
Ether	2-phenoxy-Ethanol	$C_8H_{10}O_2$	122-99-6	123	171	—
Ether	ethoxy-Acetic acid	$C_4H_8O_3$	627-03-2	123	171	—
Ether	ethoxy-Ethene	C_4H_8O	109-92-2	123	171	—
Ether	ethoxy-Ethyne	C_4H_8O	927-80-0	123	171	—
Ether	methoxy-Benzene	C_7H_8O	100-66-3	122, 212	170, 213	—
Pentanol Series	1-Pentanol	$C_5H_{12}O$	71-41-0	124	172	—
Pentanol Series	2-Pentanol	$C_5H_{12}O$	6032-29-7	124	172	—
Pentanol Series	3-Pentanol	$C_5H_{12}O$	584-02-1	124	172	—
Pentanol Series	Cyclopentanol	$C_5H_{10}O$	96-41-3	124	172	—
Pentanol Series	2-Butene-1,4-diol	$C_4H_8O_2$	110-64-5	124	172	—

Pentanol Series						
Aldehyde	2-Propene-1-ol	C$_3$H$_6$O	107-18-6	124	172	—
Aldehyde	2-Propynal	C$_3$H$_2$O	624-67-9	125	173	—
Aldehyde	Butanal	C$_4$H$_8$O	123-72-8	125, 216	173, 217	—
Aldehyde	Decanal	C$_{10}$H$_{20}$O	112-31-2	125, 216	173, 217	—
Aldehyde	Heptanal	C$_7$H$_{14}$O	111-71-7	125, 216	217	—
Aldehyde	Hexanal	C$_6$H$_{12}$O	66-25-1	—	173	—
Aldehyde	Pentanal	C$_5$H$_{10}$O	110-62-3	125	173	—
Ketone	Cyclohexanone	C$_6$H$_{10}$O	108-94-1	126, 214	174, 215	—
Ketone	2-Cyclohexen-1-one	C$_6$H$_8$O	930-68-7	126	174	—
Ketone	2-Hexanone	C$_6$H$_{12}$O	591-78-6	126	174	—
Ketone	2-Propanone (Acetone)	C$_3$H$_6$O	67-64-1	126	174	—
Ketone	3-Hexanone	C$_6$H$_{12}$O	589-38-8	118, 126	166, 174	—
Ketone	2-Pentanone	C$_5$H$_{10}$O	107-87-9	214	215	—
Ketone	2-Butanone	C$_4$H$_8$O	78-93-3	214	215	—
Four-Carbon (Butyl) Comparison	2-Butanol	C$_4$H$_{10}$O	78-92-2	127	175	—
Four-Carbon (Butyl) Comparison	2-Butanone	C$_4$H$_8$O	78-93-3	127, 214	175	—
Four-Carbon (Butyl) Comparison	2-ethoxy-Ethanol	C$_4$H$_{10}$O$_2$	110-80-5	127	175	—
Four-Carbon (Butyl) Comparison	Butanoic Acid	C$_4$H$_8$O$_2$	107-92-6	127	175	—
Four-Carbon (Butyl) Comparison	Ethoxy-Ethene	C$_4$H$_8$O	109-92-2	127		—
Four-Carbon (Butyl) Comparison	Ethoxy-Ethyne	C$_4$H$_6$O	927-80-0	127	175	—
Polyfunctional Comparison	2-Butenal	C$_4$H$_6$O	123-73-9	128	176	—
Polyfunctional Comparison	2-Furanmethanol	C$_5$H$_6$O$_2$	98-00-0	128	176	—
Polyfunctional Comparison	3-methyl-2-Butenal	C$_5$H$_8$O	107-86-8	128	176	—
Polyfunctional Comparison	3-phenyl-2-Propenal	C$_9$H$_8$O	14371-10-9	128	176	—
Polyfunctional Comparison	Benzenepropanal	C$_9$H$_{10}$O	104-53-0	128	176	—
Aldehyde, Substituted Aromatic	2,4,6-trimethyl-Benzaldehyde	C$_{10}$H$_{12}$O	487-68-3	129	177	—
Aldehyde, Substituted Aromatic	2,4-dimethyl-Benzaldehyde	C$_9$H$_{10}$O	15764-16-6	129	177	—
Aldehyde, Substituted Aromatic	2-bromo-Benzaldehyde	C$_7$H$_5$BrO	6630-33-7	129	177	—
Aldehyde, Substituted Aromatic	2-chloro-Benzaldehyde	C$_7$H$_5$ClO	89-98-5	129	177	—
Aldehyde, Substituted Aromatic	Benzaldehyde	C$_7$H$_6$O	100-52-7	129	177	—
Alkane	Methylcyclopentane	C$_6$H$_{12}$	96-37-7	132		—
Alkane	Cycloheptane	C$_7$H$_{14}$	291-64-5	132		—
Alkane	Cyclohexane	C$_6$H$_{12}$	110-82-7	132, 218		—

(continued)

Functional Group	Compound Name	Molecular Formula	CAS Number	$10,500 - 6300$ cm^{-1} ($952 - 1587$ nm)	$7200-3800$ cm^{-1} ($1389 - 2632$ nm)	Other NIR Regions
Alkane	Cyclopentane	C_5H_{10}	287-92-3	132	—	—
Alkane	Dodecane	$C_{12}H_{26}$	112-40-3	130	178	—
Alkane	Heptane	C_7H_{16}	142-82-5	130	178	—
Alkane	Hexane	C_6H_{14}	110-54-3	130	—	—
Alkane	Octadecane	$C_{18}H_{38}$	593-45-3	130	178	—
Alkane	Octane	C_8H_{18}	111-65-9	130	178	—
Alkane	Pentane	C_5H_{12}	109-66-0	130	178	—
Alkane	Undecane	$C_{11}H_{24}$	1120-21-4	118, 234, 235	166, 236, 237	—
Butyl Comparison Group	1-Butanol	$C_4H_{10}O$	71-36-3	131	179	—
Butyl Comparison Group	1-chlorobutane	C_4H_9Cl	109-69-3	131	179	—
Butyl Comparison Group	2-bromo-Butane	C_4H_9Br	78-76-2	131	179	—
Butyl Comparison Group	2-Butanol	$C_4H_{10}O$	78-92-2	131	179	—
Butyl Comparison Group	2-chloro-Butane	C_4H_9Cl	78-86-4	131	179	—
Butyl Comparison Group	2-iodo-Butane	C_4H_9I	513-48-4	131	179	—
Alkane & Cycloalkane Comparison	Cycloheptane	C_7H_{14}	291-64-5	132	180	—
Alkane & Cycloalkane Comparison	Cyclohexane	C_6H_{12}	110-82-7	132	180	—
Alkane & Cycloalkane Comparison	Cyclopentane	C_5H_{10}	287-92-3	132	180	—
Alkane & Cycloalkane Comparison	methyl-Cyclopentane	C_6H_{12}	96-37-7	132	180	—
Ethyl (C_2) Comparison Group	1,1,1-trichloro-Ethane	$C_2H_4Cl_2$	71-55-6	133	181	—
Ethyl (C_2) Comparison Group	1,2-dibromo-Ethane	$C_2H_4Br_2$	106-93-4	133	181	—
Ethyl (C_2) Comparison Group	1,2-dichloro-Ethane	$C_2H_4Cl_2$	107-06-2	133	181	—
Ethyl (C_2) Comparison Group	2-bromo-Ethanol	C_2H_5BrO	540-51-2	133	181	—
Ethyl (C_2) Comparison Group	nitro-Ethane	$C_2H_5NO_2$	79-24-3	133	181	—
Isomerism Comparison Group	2,2,4-trimethylpentane	C_8H_{18}	540-84-1	134	182	—
Isomerism Comparison Group	2-methylpentane	C_6H_{14}	107-83-5	134	182	—
Isomerism Comparison Group	Heptane	C_7H_{16}	142-82-5	134	182	—
Isomerism Comparison Group	Pentane	C_5H_{12}	109-66-0	134	182	—
Methyl Comparison Group	dichloromethyl Silane	CH_4Cl_2	75-54-7	135	183	—

Group	Name	Formula	CAS			
Methyl Comparison Group	Formic acid, methyl ester	$C_2H_4O_2$	107-31-3	135	183	—
Methyl Comparison Group	Methanol	CH_4O	67-56-1	135	183	—
Methyl Comparison Group	nitro-Methane	CH_3NO_2	75-52-5	135	183	—
Propyl Comparison Group	1-chloro-Propane	C_3H_7Cl	540-54-5	136	184	—
Propyl Comparison Group	1-Propanol	C_3H_8O	71-23-8	136	184	—
Propyl Comparison Group	2-chloro Propane	C_3H_7Cl	75-29-6	136	184	—
Propyl Comparison Group	2-Propanol	C_3H_8O	67-63-0	136	184	—
Cycloalkene	Cycloheptene	C_7H_{12}	628-92-2	137	186	—
Cycloalkene	Cyclohexene	C_6H_{10}	110-83-8	137	186	—
Cycloalkene	Cyclooctene	C_8H_{14}	931-87-3	137	186	—
Cycloalkene	Cyclopentene	C_5H_8	142-29-0	137	186	—
Alkene	1-Decene	$C_{10}H_{20}$	872-05-9	138, 222	185, 223	—
Alkene	1-Heptene	C_7H_{14}	592-76-7	222	223	—
Alkene	1-Hexene	C_6H_{12}	592-41-6	138	185	—
Alkene	1-Nonene	C_9H_{18}	124-11-8	222	223	—
Alkene	1-Octene	C_8H_{16}	111-66-0	138	185	—
Alkene	1-Pentene	C_5H_{10}	109-67-1	138	185	—
Alkene	1-Tetradecene	$C_{14}H_{28}$	1120-36-1	138	185	—
Alkene, Internal versus Terminal	2,3-dimethyl-2-Butene	C_6H_{12}	563-79-1	139	187	—
Alkene, Internal versus Terminal	2-methyl-2-Butene	C_5H_{10}	513-35-9	139	187	—
Alkene, Internal versus Terminal	3,3-dimethyl-1-Butene	C_6H_{12}	558-37-2	139	187	—
Alkyne	1-Heptyne	C_7H_{12}	628-71-7	226	227	—
Alkyne	1-Octyne	C_8H_{14}	629-05-0	140, 226	188, 227	—
Alkyne	1-Pentyne	C_5H_8	627-19-0	140, 226	188, 227	—
Alkyne	2-Butyne	C_4H_6	503-17-3	140	188	—
Alkyne	3-chloro-1-Propyne	C_3H_3Cl	624-65-7	140	188	—
Alkyne	4-Octyne	C_8H_{14}	1942-45-6	140	188	—
X-H Compound Comparison	1-Butanamine (C-NH)	$C_4H_{11}N$	109-73-9	141	189	—
X-H Compound Comparison	2-Butanethiol (C-SH)	$C_4H_{10}S$	513-53-1	141	189	—
X-H Compound Comparison	2-Butanol (C-OH)	$C_4H_{10}O$	78-92-2	141	189	—
X-H Compound Comparison	2-Propanamine (C-NH)	C_3H_9N	75-31-0	141	189	—
X-H Compound Comparison	Pentane (C-C-H)	C_5H_{12}	109-66-0	141	189	—
Alkyl Amine	1,2-Propanediamine	$C_3H_{10}N_2$	78-90-0	142, 224	190, 225	—
Alkyl Amine	1,7-Heptanediamine	$C_7H_{18}N_2$	646-19-5	142	190	—
Alkyl Amine	1-Decanamine	$C_{10}H_{23}N$	2016-57-1	118, 142, 224, 234, 235	166, 190, 225, 236, 237	—

(continued)

Functional Group	Compound Name	Molecular Formula	CAS Number	10,500 – 6300 cm⁻¹ (952 – 1587 nm)	7200-3800 cm⁻¹ (1389 – 2632 nm)	Other NIR Regions
Alkyl Amine	Cyclododecanamine	$C_{12}H_{25}N$	1502-03-0	142, 224	190, 225	—
Alkyl Amine	N,N,N´,N´-tetramethyl-1,4-Butanediamine	$C_8H_{20}N_2$	111-51-3	142	190	—
Alkyl Amine	N,N-dipropyl-1-Propanamine	$C_9H_{21}N$	102-69-2	142	190	—
Aryl Amine	2-Pyridinemethanol	C_6H_7NO	586-98-1	143	191	—
Aryl Amine	4-methyl-Pyridine	C_6H_7N	108-89-4	143	191	—
Aryl Amine	Benzenamine	C_7H_7N	62-53-3	143	191	—
Aryl Amine	Indole, 1-H	C_8H_7N	120-72-9	143	191	—
Aryl Amine	N,N-dimethyl-1,3-Benzenediamine	$C_8H_{12}N_2$	2836-04-2	143	191	—
Amide	N-methyl-Acetamide	C_3HH_7NO	79-16-3	144, 145, 228	192, 193, 229	—
Amide	Formamide	CH_3NO	75-12-7	144, 228	192, 229	—
Amide	hexamethyl-Phosphoric triamide	$C_6H_{18}N_3OP$	680-31-9	144	192	—
Amide	N,N-dimethyl-2-Propenamide	C_5H_9NO	2680-03-7	144	192	—
Amide	N-methyl Propanamide	C_4H_9NO	187-58-2	145	193	—
Amide	N, N-dimethyl-3-methyl-Benzamide	$C_{10}H_{13}NO$	—	145	193	—
Amide	N, N-dimethyl-3-oxo-Benzamide	$C_9H_{10}NO_2$	—	145	193	—
Amide	N-methyl Formamide	C_2H_5NO	123-39-7	145	193	—
Amino Acid	D-Valine, 3-methyl, 1,1-dimethylethyl ester	$C_{10}H_{21}NO_2$	61169-85-5	146	194	—
Amino Acid	L-Valine, 3-methyl-, 1,1-dimethylethyl ester	$C_{10}H_{21}NO_2$	31556-74-8	146	194	—
Amino Acid	N-formyl-Glycine, ethyl ester	$C_5H_9NO_3$	4172-32-1	146	194	—
Aromatic Compound with N or S	1-methyl-1H-Pyrrole	C_5H_7N	96-54-8	147	195	—
Aromatic Compound with N or S	4-methyl-Pyridine	C_6H_7N	108-89-4	147	195	—
Aromatic Compound with N or S	Indole, 1H-	C_8H_7N	120-72-9	147	195	—

Group	Compound	Formula	CAS			
Aromatic Compound with N or S	Quinoline	C_9H_7N	91-22-5	147	195	—
Aromatic Compound with N or S	Thiophene	C_4H_4S	110-02-1	147	195	—
Aryl versus Alkyl Series	1,7-Octadiene	C_8H_{14}	3710-30-3	148	196	—
Aryl versus Alkyl Series	2,2,4-trimethyl-Pentane	C_8H_{18}	540-84-1	118, 148, 218, 219, 234, 235	166, 196, 220, 221, 236, 237	—
Aryl versus Alkyl Series	Benzaldehyde	C_7H_6O	100-52-7	149	197	—
Aryl versus Alkyl Series	Benzene	C_6H_6	71-43-2	149, 218, 219	197, 220, 221	—
Aryl versus Alkyl Series	Cyclohexane	C_6H_{12}	110-82-7	148, 218, 219	196, 220, 221	—
Aryl versus Alkyl Series	Cyclohexene	C_6H_{10}	110-83-8	148	196	—
Aryl versus Alkyl Series	methyl-Benzene	C_7H_8	108-88-3	149, 234, 235	166, 197, 236, 237	—
Aryl versus Alkyl Series	Octane	C_8H_{18}	111-65-9	148, 218, 219	196, 220, 221	—
Aromatic (Aryl) Compound	1,1-Biphenyl	$C_{12}H_{10}$	92-52-4	150	198	—
Aromatic (Aryl) Compound	1,2-dimethyl-Benzene (xylene)	C_8H_{10}	95-47-6	150	—	—
Aromatic (Aryl) Compound	1,3-dimethyl-Benzene (xylene)	C_8H_{10}	108-38-3	150	198	—
Aromatic (Aryl) Compound	1,4-dimethyl-Benzene (xylene)	C_8H_{10}	106-42-3	150	198	—
Aromatic (Aryl) Compound	2,4,5-trichloro-Phenol	$C_6H_3Cl_3O$	95-95-4	150	198	—
Aromatic (Aryl) Compound	2-ethyl-Phenol	$C_8H_{10}O$	90-00-6	150	198	—
Aromatic (Aryl) Compound	3-methyl-Phenol	C_7H_8O	08-39-4	150	198	—
Aromatic (Aryl) Compound	4-methyl-Phenol	C_7H_8O	106-44-5	151	199	—
Aromatic (Aryl) Compound	bromo-Benzene	C_6H_5Br	108-86-1	151	199	—
Aromatic (Aryl) Compound	methyl-Benzene	C_7H_8	108-88-3	118, 151	166, 199	—
Aromatic (Aryl) Compound	Naphthalene	$C_{10}H_8$	91-20-3	151	199	—
Aromatic (Aryl) Compound	nitro-Benzene	$C_6H_5NO_2$	98-95-3	151	199	—
Aromatic (Aryl) Compound	Phenanthrene	$C_{14}H_{10}$	85-01-8	151	199	—
Cyclic Ether	3,4-dihydro-2H-Pyran	C_5H_8O	110-87-2	152	200	—
Cyclic Ether	Furan	C_4H_4O	110-00-9	152	200	—
Cyclic Ether	tetrahydro-2H-Pyran	$C_5H_{10}O$	142-68-7	152	200	—
Cyclic Ether	tetrahydro-4H-Pyran-4-one	$C_5H_8O_2$	29943-42-8	152	200	—
Amides, Comparison Group	N,N-diethyl-3-methyl-Benzamide	$C_{12}H_{17}NO$	134-62-3		193	—
Amides, Comparison Group	N,N-dimethyl-3-oxo-Butanamide	$C_8H_{15}NO_2$	2235-46-3		193	—
Amides, Comparison Group	N-methyl-Acetamide	C_3H_7NO	79-16-3	153	193, 201	—

(continued)

Functional Group	Compound Name	Molecular Formula	CAS Number	10,500 – 6300 cm^{-1} (952 – 1587 nm)	7200-3800 cm^{-1} (1389 – 2632 nm)	Other NIR Regions
Amides, Comparison Group	N-methyl-Formamide	C$_2$H$_5$NO	123-39-7	—	193	—
Amides, Comparison Group	N-methyl-Propanamide	C$_4$H$_9$NO	187-58-2	—	193	—
Acid vs. Amide vs. Ester Comparison	1-Propen-2-ol acetate	C$_5$H$_8$O$_2$	108-22-5	153	201	—
Acid vs. Amide vs. Ester Comparison	2-ethoxy- acetate-Ethanol	C$_6$H$_{12}$O$_3$	111-15-9	153	201	—
Acid vs. Amide vs. Ester Comparison	Acetic acid	C$_4$H$_4$O$_2$	64-19-7	153	201	—
Acid vs. Amide vs. Ester Comparison	N-methyl-Acetamide	C$_3$H$_7$NO	79-16-3	153	—	—
Carboxylic Acid	Acetic acid	C$_4$H$_4$O$_2$	64-19-7	230	231	—
Carboxylic Acid	Benzoic acid anhydride	C$_{14}$H$_{10}$O$_3$	93-97-0	230	231	—
Carboxylic Acid	Decanoic acid	C$_{10}$H$_{20}$O$_2$	334-48-5	—	231	—
Carboxylic Acid	Dodecanoic acid	C$_{12}$H$_{24}$O$_2$	143-07-7	154, 230	202	—
Carboxylic Acid	Formic acid	CH$_2$O$_2$	64-18-6	154, 230	202, 231	—
Carboxylic Acid	Methyl ester-Benzoic acid	C$_8$H$_8$O$_2$	93-58-3	154	202	—
Carboxylic Acid	Pentanoic acid	C$_5$H$_{10}$O$_2$	109-52-4	—	202	—
Carboxylic Acid	Propanoic acid	C$_3$H$_6$O$_2$	79-09-4	154	—	—
Heterocyclic Compound	1H-Indole	C$_8$H$_7$N	120-72-9	155	203	—
Heterocyclic Compound	1-methyl-1H-Pyrrole	C$_5$H$_7$N	96-54-8	155	203	—
Heterocyclic Compound	2-bromo-Pyridine	C$_5$H$_4$BrN	109-04-6	155	203	—
Heterocyclic Compound	3-ethyl-Pyridine	C$_7$H$_9$N	536-78-7	155	203	—
Heterocyclic Compound	Furan	C$_4$H$_4$O	110-00-9	155	203	—
Heterocyclic Compound	Quinoline	C$_9$H$_7$N	91-22-5	155	203	—
Heterocyclic Compound	Thiophene	C$_4$H$_4$S	110-02-1	155	203	—
Heterocyclic (5-member ring)	2,2-dimethyl-Thiazolidine	C$_5$H$_{11}$NS	19351-18-9	156	204	—
Heterocyclic (5-member ring)	4,5-dihydro-2-(2-propenylthio)-Thiazole	C$_6$H$_9$NS$_2$	3571-74-2	156	204	—
Heterocyclic (5-member ring)	4,5-dihydro-2(methylthio)-Thiazole	C$_4$H$_7$NS$_2$	19975-56-5	156	204	—
Functional Group Comparisons	Undecane	C$_{11}$H$_{24}$	1120-21-4	157	205	—
Functional Group Comparisons	1-Decanamine	C$_{10}$H$_{23}$N	2016-57-1	157, 234	205	—

Group	Name	Formula	CAS			
Functional Group Comparisons	2,2,4-trimethyl-Pentane	C_8H_{18}	540-84-1	157	205	—
Functional Group Comparisons	2-methyl-2-Propanol	$C_4H_{10}O$	75-65-0	157, 234, 235	205, 236, 237	—
Functional Group Comparisons	3-Hexanone	$C_6H_{12}O$	589-38-8	157, 234, 235	205, 236, 237	—
Functional Group Comparisons	Benzene	C_6H_6	71-43-2	157	205	—
Functional Group Comparisons	methyl-Benzene	C_7H_8	108-88-3	157	205	—
Polyene	1,3,5-Cycloheptatriene	C_7H_8	544-25-2	158	206	—
Polyene	1,3-Cyclohexadiene	C_6H_8	592-57-4	158	206	—
Polyene	1,3-Cyclooctadiene	C_8H_{12}	1700-10-3	158	206	—
Polyene	1,5-Hexadiene	C_6H_{10}	592-42-7	158, 232	206, 233	—
Polyene	1,7-Octadiene	C_8H_{14}	3710-30-3	232	233	—
Polyene	2-methyl-1,3-Butadiene	C_5H_8	78-79-5	158	206	—
Polymers and Rubbers	Ethylene Vinyl Acetate	polymer	N/A	—	—	159
Polymers and Rubbers	60% polypropylene and 40% polyester	polymer	N/A	—	—	159
Polymers and Rubbers	60% polypropylene-polyethylene acopolymer and 40% polyester	polymer	N/A	—	—	159
Polymers and Rubbers	Atactic Polypropylene	polymer	N/A	—	—	159
Polymers and Rubbers	Styrene isoprene styrene	polymer	N/A	—	—	160
Polymers and Rubbers	Polyacrylic acid	polymer	N/A	—	—	160
Polymers and Rubbers	Polystyrene	polymer	N/A	—	—	160
Polymers and Rubbers	Silicone	polymer	N/A	—	—	160
Polymers and Rubbers	Starch	polymer	N/A	—	—	160
Polymers and Rubbers	Styrene, ethylene, styrene copolymer	polymer	N/A	—	—	160
Polymers and Rubbers	Polypropylene, isotactic, chlorinated	polymer	N/A	—	—	161
Polymers and Rubbers	Cellulose acetate	polymer	N/A	—	—	161
Polymers and Rubbers	Cellulose acetate butyrate	polymer	N/A	—	—	161
Polymers and Rubbers	Ethyl cellulose	polymer	N/A	—	—	161
Polymers and Rubbers	Polyethylene, chlorinated (25% Cl)	polymer	N/A	—	—	161
Polymers and Rubbers	Poly(ethylene oxide)	polymer	N/A	—	—	161
Polymers and Rubbers	Poly(isobutyl methacrylate)	polymer	N/A	—	—	161
Polymers and Rubbers	Poly(vinylidene fluoride)	polymer	N/A	—	—	162

(continued)

Functional Group	Compound Name	Molecular Formula	CAS Number	10,500 – 6300 cm^{-1} (952 – 1587 nm)	7200-3800 cm^{-1} (1389 – 2632 nm)	Other NIR Regions
Polymers and Rubbers	Cellulose propionate	polymer	N/A	—	—	162
Polymers and Rubbers	Poly(vinyl butyral)	polymer	N/A	—	—	162
Polymers and Rubbers	Poly(vinyl chloride)	polymer	N/A	—	—	162
Polymers and Rubbers	Poly(vinyl pyrrolidone)	polymer	N/A	—	—	162
Polymers and Rubbers	Poly(vinyl stearate)	polymer	N/A	—	—	162
Third Overtone C-H Comparison	Toluene	C$_7$H$_8$	108-88-3	—	—	163, 238
Third Overtone C-H Comparison	Trimethyl pentane	C$_8$H$_{18}$	540-84-1	—	—	163, 238
Third Overtone C-H Comparison	n-Decane	C$_{10}$H$_{22}$	124-18-5	—	—	163, 238
Third Overtone C-H Comparison	Acetone	C$_3$C$_6$O	67-64-1	—	—	163, 238
Third Overtone C-H Comparison	tert-Butanol	C$_4$H$_{10}$O	75-65-0	—	—	163, 238

Note: N/A indicates natural or synthetic product with no designated CAS number.

Appendix 6: Spectra Index by Alphanumerical Order of Compound Name

Functional Group	Compound Name	Molecular Formula	CAS Number	10,500 – 6300 cm⁻¹ (952 – 1587 nm)	7200-3800 cm⁻¹ (1389 – 2632 nm)	Other NIR Regions
Ethyl (C₂) Comparison Group	1,1,1-trichloro-Ethane	$C_2H_4Cl_2$	71-55-6	133	181	—
Ether	1,1,2-trimethoxy-Ethane	$C_5H_{12}O_3$	24332-20-5	122, 212	170, 213	—
Aromatic (Aryl) Compound	1,1-Biphenyl	$C_{12}H_{10}$	92-52-4	150	198	—
Ether	1,1-dimethoxy-Ethane	$C_4H_{10}O_2$	534-15-6	122	170	—
Alcohol (diol)	1,2 Propanediol	$C_3H_8O_2$	57-55-6	120	168	—
Butanediol Series	1,2-Butanediol	$C_4H_{10}O_2$	584-03-2	121	169, 211	—
Ethyl (C₂) Comparison Group	1,2-dibromo-Ethane	$C_2H_4Br_2$	106-93-4	133	181	—
Ethyl (C₂) Comparison Group	1,2-dichloro-Ethane	$C_2H_4Cl_2$	107-06-2	133	181	—
Ether	1,2-diethoxy-Ethane	$C_6H_{14}O_2$	629-14-1	122	170	—
Aromatic (Aryl) Compound	1,2-dimethyl-Benzene (xylene)	C_8H_{10}	95-47-6	150	—	—
Alcohol (diol)	1,2-Ethanediol	$C_2H_6O_2$	107-21-1	120	168	—
Alkyl Amine	1,2-Propanediamine	$C_3H_{10}N_2$	78-90-0	142, 224	190, 225	—
Polyene	1,3,5-Cycloheptatriene	C_7H_8	544-25-2	158	206	—
Butanediol Series	1,3-Butanediol	$C_4H_{10}O_2$	107-88-0	121	169, 211	—
Polyene	1,3-Cyclohexadiene	C_6H_8	592-57-4	158	206	—
Polyene	1,3-Cyclooctadiene	C_8H_{12}	1700-10-3	158	206	—
Aromatic (Aryl) Compound	1,3-dimethyl-Benzene (xylene)	C_8H_{10}	108-38-3	150	198	—
Butanediol Series	1,4-Butanediol	$C_4H_{10}O_2$	110-63-4	121	169, 211	—
Aromatic (Aryl) Compound	1,4-dimethyl-Benzene (xylene)	C_8H_{10}	106-42-3	150	198	—
Polyene	1,5-Hexadiene	C_6H_{10}	592-42-7	158, 232	206, 233	—
Alkyl Amine	1,7-Heptanediamine	$C_7H_{18}N_2$	646-19-5	142	190	—
Aryl versus Alkyl Series	1,7-Octadiene	C_8H_{14}	3710-30-3	148	196	—
Polyene	1,7-Octadiene	C_8H_{14}	3710-30-3	232	233	—
X-H Compound Comparison	1-Butanamine (C-NH)	$C_4H_{11}N$	109-73-9	141	189	—
Alcohol (mono-OH)	1-Butanol	$C_4H_{10}O$	71-36-3	119, 208, 209	167, 210	—
Butyl Comparison Group	1-Butanol	$C_4H_{10}O$	71-36-3	131	179	—
Butyl Comparison Group	1-chlorobutane	C_4H_9Cl	109-69-3	131	179	—
Propyl Comparison Group	1-chloro-Propane	C_3H_7Cl	540-54-5	136	184	—
Alkyl Amine	1-Decanamine	$C_{10}H_{23}N$	2016-57-1	118, 142, 224, 234, 235	166, 190, 225, 236, 237	—
Functional Group Comparisons	1-Decanamine	$C_{10}H_{23}N$	2016-57-1	157, 234	205	—
Alkene	1-Decene	$C_{10}H_{20}$	872-05-9	138, 222	185, 223	—

Alkene	1-Heptene	C$_7$H$_{14}$	592-76-7	222	223	—
Alkyne	1-Heptyne	C$_7$H$_{12}$	628-71-7	226	227	—
Alkene	1-Hexene	C$_6$H$_{12}$	592-41-6	138	185	—
Heterocyclic Compound	1H-Indole	C$_8$H$_7$N	120-72-9	155	203	—
Aromatic Compound with N or S	1-methyl-1H-Pyrrole	C$_5$H$_7$N	96-54-8	147	195	—
Heterocyclic Compound	1-methyl-1H-Pyrrole	C$_5$H$_7$N	96-54-8	155	203	—
Alkene	1-Nonene	C$_9$H$_{18}$	124-11-8	222	223	—
Alkene	1-Octene	C$_8$H$_{16}$	111-66-0	138	185	—
Alkyne	1-Octyne	C$_8$H$_{14}$	629-05-0	140, 226	188, 227	—
Pentanol Series	1-Pentanol	C$_5$H$_{12}$O	71-41-0	124	172	—
Alkene	1-Pentene	C$_5$H$_{10}$	109-67-1	138	185	—
Alkyne	1-Pentyne	C$_5$H$_8$	627-19-0	140, 226	188, 227	—
Alcohol (mono-OH)	1-Propanol	C$_3$H$_8$O	71-23-8	119,209	—	—
Propyl Comparison Group	1-Propanol	C$_3$H$_8$O	71-23-8	136	184	—
Acid vs. Amide vs. Ester Comparison	1-Propen-2-ol acetate	C$_5$H$_8$O$_2$	108-22-5	153	201	—
Alkene	1-Tetradecene	C$_{14}$H$_{28}$	1120-36-1	138	185	—
Isomerism Comparison Group	2,2,4-trimethylpentane	C$_8$H$_{18}$	540-84-1	134	182	—
Aryl versus Alkyl Series	2,2,4-trimethyl-Pentane	C$_8$H$_{18}$	540-84-1	118, 148, 218, 219, 234, 235	166, 196, 220, 221, 236, 237	—
Functional Group Comparisons	2,2,4-trimethyl-Pentane	C$_8$H$_{18}$	540-84-1	157	205	—
Ether	2,2-dimethoxy-Propane	C$_5$H$_{12}$O$_2$	77-76-7	122, 212	170, 213	—
Heterocyclic (5-member ring)	2,2-dimethyl-Thiazolidine	C$_5$H$_{11}$NS	19351-18-9	156	204	—
Alcohol (diol)	2,3-Butanediol	C$_4$H$_{10}$O$_2$	513-85-9	120, 210	168, 211	—
Butanediol Series	2,3-Butanediol	C$_4$H$_{10}$O$_2$	513-85-9	121, 208	169	—
Alkene, Internal versus Terminal	2,3-dimethyl-2-Butene	C$_6$H$_{12}$	563-79-1	139	187	—
Aromatic (Aryl) Compound	2,4,5-trichloro-Phenol	C$_6$H$_3$Cl$_3$O	95-95-4	150	198	—
Aldehyde, Substituted Aromatic	2,4,6-trimethyl-Benzaldehyde	C$_{10}$H$_{12}$O	487-68-3	129	177	—
Aldehyde, Substituted Aromatic	2,4-dimethyl-Benzaldehyde	C$_9$H$_{10}$O	15764-16-6	129	177	—
Alcohol (diol)	2,4-Pentanediol	C$_5$H$_{12}$O$_2$	625-69-4	120	168	—
Alcohol (diol)	2,5-Hexanediol	C$_6$H$_{14}$O$_2$	2935-44-6	120	168	—
Aldehyde, Substituted Aromatic	2-bromo-Benzaldehyde	C$_7$H$_5$BrO	6630-33-7	129	177	—
Butyl Comparison Group	2-bromo-Butane	C$_4$H$_9$Br	78-76-2	131	179	—
Ethyl (C$_2$) Comparison Group	2-bromo-Ethanol	C$_2$H$_5$BrO	540-51-2	133	181	—
Heterocyclic Compound	2-bromo-Pyridine	C$_5$H$_4$BrN	109-04-6	155	203	—
X-H Compound Comparison	2-Butanethiol (C-SH)	C$_4$H$_{10}$S	513-53-1	141	189	—

(continued)

Functional Group	Compound Name	Molecular Formula	CAS Number	10,500 – 6300 cm⁻¹ (952 – 1587 nm)	7200-3800 cm⁻¹ (1389 – 2632 nm)	Other NIR Regions
Alcohol (mono-OH)	2-Butanol	$C_4H_{10}O$	78-92-2	209	—	—
Four-Carbon (Butyl) Comparison	2-Butanol	$C_4H_{10}O$	78-92-2	127	175	—
Butyl Comparison Group	2-Butanol	$C_4H_{10}O$	78-92-2	131	179	—
X-H Compound Comparison	2-Butanol (C-OH)	$C_4H_{10}O$	78-92-2	141	189	—
Ketone	2-Butanone	C_4H_8O	78-93-3	214	215	—
Four-Carbon (Butyl) Comparison	2-Butanone	C_4H_8O	78-93-3	127, 214	175	—
Polyfunctional Comparison	2-Butenal	C_4H_6O	123-73-9	128	176	—
Pentanol Series	2-Butene-1,4-diol	$C_4H_8O_2$	110-64-5	124	172	—
Alkyne	2-Butyne	C_4H_6	503-17-3	140	188	—
Propyl Comparison Group	2-chloro Propane	C_3H_7Cl	75-29-6	136	184	—
Aldehyde, Substituted Aromatic	2-chloro-Benzaldehyde	C_7H_5ClO	89-98-5	129	177	—
Butyl Comparison Group	2-chloro-Butane	C_4H_9Cl	78-86-4	131	179	—
Ketone	2-Cyclohexen-1-one	C_6H_8O	930-68-7	126	174	—
Acid vs. Amide vs. Ester Comparison	2-ethoxy- acetate-Ethanol	$C_6H_{12}O_3$	111-15-9	153	201	—
Ether	2-ethoxy-Ethanol	$C_4H_{10}O_2$	110-80-5	123	171	—
Four-Carbon (Butyl) Comparison	2-ethoxy-Ethanol	$C_4H_{10}O_2$	110-80-5	127	175	—
Aromatic (Aryl) Compound	2-ethyl-Phenol	$C_8H_{10}O$	90-00-6	150	198	—
Polyfunctional Comparison	2-Furanmethanol	$C_5H_6O_2$	98-00-0	128	176	—
Ketone	2-Hexanone	$C_6H_{12}O$	591-78-6	126	174	—
Butyl Comparison Group	2-iodo-Butane	C_4H_9I	513-48-4	131	179	—
Polyene	2-methyl-1,3-Butadiene	C_5H_8	78-79-5	158	206	—
Alkene, Internal versus Terminal	2-methyl-2-Butene	C_5H_{10}	513-35-9	139	187	—
Alcohol (mono-OH)	2-methyl-2-Propanol	$C_4H_{10}O$	75-65-0	118, 119, 209, 234, 235	166,167, 236, 237	—
Functional Group Comparisons	2-methyl-2-Propanol	$C_4H_{10}O$	75-65-0	157, 234, 235	205, 236, 237	—
Isomerism Comparison Group	2-methylpentane	C_6H_{14}	107-83-5	134	182	—
Pentanol Series	2-Pentanol	$C_5H_{12}O$	6032-29-7	124	172	—
Ketone	2-Pentanone	$C_5H_{10}O$	107-87-9	214	215	—
Ether	2-phenoxy-Ethanol	$C_8H_{10}O_2$	122-99-6	123	171	—
X-H Compound Comparison	2-Propanamine (C-NH)	C_3H_9N	75-31-0	141	189	—
Alcohol (mono-OH)	2-Propanol	C_3H_8O	67-63-0	119, 209	167	—
Propyl Comparison Group	2-Propanol	C_3H_8O	67-63-0	136	184	—

Ketone	2-Propanone (Acetone)	C$_3$H$_6$O	67-64-1	126	174	—
Pentanol Series	2-Propene-1-ol	C$_3$H$_6$O	107-18-6	124	172	—
Aldehyde	2-Propynal	C$_3$H$_2$O	624-67-9	125	173	—
Aryl Amine	2-Pyridinemethanol	C$_6$H$_7$NO	586-98-1	143	191	—
Alkene, Internal versus Terminal	3,3-dimethyl-1-Butene	C$_6$H$_{12}$	558-37-2	139	187	—
Cyclic Ether	3,4-dihydro-2H-Pyran	C$_5$H$_8$O	110-87-2	152	200	—
Alkyne	3-chloro-1-Propyne	C$_3$H$_3$Cl	624-65-7	140	188	—
Heterocyclic Compound	3-ethyl-Pyridine	C$_7$H$_9$N	536-78-7	155	203	—
Ketone	3-Hexanone	C$_6$H$_{12}$O	589-38-8	118, 126	166, 174	—
Functional Group Comparisons	3-Hexanone	C$_6$H$_{12}$O	589-38-8	157, 234, 235	205, 236, 237	—
Polyfunctional Comparison	3-methyl-2-Butenal	C$_5$H$_8$O	107-86-8	128	176	—
Aromatic (Aryl) Compound	3-methyl-Phenol	C$_7$H$_8$O	08-39-4	150	198	—
Alcohol (mono-OH)	3-Pentanol	C$_5$H$_{12}$O	584-02-01	119, 209	167	—
Pentanol Series	3-Pentanol	C$_5$H$_{12}$O	584-02-1	124	172	—
Polyfunctional Comparison	3-phenyl-2-Propenal	C$_9$H$_8$O	14371-10-9	128	176	—
Heterocyclic (5-member ring)	4,5-dihydro-2-(2-propenylthio)-Thiazole	C$_6$H$_9$NS$_2$	3571-74-2	156	204	—
Heterocyclic (5-member ring)	4,5-dihydro-2(methylthio)-Thiazole	C$_4$H$_7$NS$_2$	19975-56-5	156	204	—
Aromatic (Aryl) Compound	4-methyl-Phenol	C$_7$H$_8$O	106-44-5	151	199	—
Aryl Amine	4-methyl-Pyridine	C$_6$H$_7$N	108-89-4	143	191	—
Aromatic Compound with N or S	4-methyl-Pyridine	C$_6$H$_7$N	108-89-4	147	195	—
Alkyne	4-Octyne	C$_8$H$_{14}$	1942-45-6	140	188	—
Polymers and Rubbers	60% polypropylene and 40% polyester	polymer	N/A		—	159
Polymers and Rubbers	60% polypropylene-polyethylene acopolymer and 40% polyester	polymer	N/A		—	159
Acid vs. Amide vs. Ester Comparison	Acetic acid	C$_2$H$_4$O$_2$	64-19-7	153	201	—
Carboxylic Acid	Acetic acid	C$_2$H$_4$O$_2$	64-19-7	230	231	—
Third Overtone C–H Comparison	Acetone	C$_3$H$_6$O	67-64-1		—	163, 238
Polymers and Rubbers	Atactic Polypropylene	polymer	N/A		—	159
Aldehyde, Substituted Aromatic	Benzaldehyde	C$_7$H$_6$O	100-52-7	129	177	—
Aryl versus Alkyl Series	Benzaldehyde	C$_7$H$_6$O	100-52-7	149	197	—
Aryl Amine	Benzenamine	C$_6$H$_7$N	62-53-3	143	191	—

(continued)

Functional Group	Compound Name	Molecular Formula	CAS Number	10,500 – 6300 cm⁻¹ (952 – 1587 nm)	7200-3800 cm⁻¹ (1389 – 2632 nm)	Other NIR Regions
Aryl versus Alkyl Series	Benzene	C_6H_6	71-43-2	149, 218, 219	197, 220, 221	—
Functional Group Comparisons	Benzene	C_6H_6	71-43-2	157	205	—
Polyfunctional Comparison	Benzenepropanal	$C_9H_{10}O$	104-53-0	128	176	—
Carboxylic Acid	Benzoic acid anhydride	$C_{14}H_{10}O_3$	93-97-0	230	231	—
Aromatic (Aryl) Compound	bromo-Benzene	C_6H_5Br	108-86-1	151	199	—
Aldehyde	Butanal	C_4H_8O	123-72-8	125, 216	173, 217	—
Four-Carbon (Butyl) Comparison	Butanoic Acid	$C_4H_8O_2$	107-92-6	127	175	—
Polymers and Rubbers	Cellulose acetate	polymer	N/A	—		161
Polymers and Rubbers	Cellulose acetate butyrate	polymer	N/A	—		161
Polymers and Rubbers	Cellulose propionate	polymer	N/A	—		162
Alkyl Amine	Cyclododecanamine	$C_{12}H_{25}N$	1502-03-0	142, 224	190, 225	—
Alkane	Cycloheptane	C_7H_{14}	291-64-5	132	—	—
Alkane & Cycloalkane Comparison	Cycloheptane	C_7H_{14}	291-64-5	132	180	—
Cycloalkene	Cycloheptene	C_7H_{12}	628-92-2	137	186	—
Alkane	Cyclohexane	C_6H_{12}	110-82-7	132, 218	—	—
Alkane & Cycloalkane Comparison	Cyclohexane	C_6H_{12}	110-82-7	132	180	—
Aryl versus Alkyl Series	Cyclohexane	C_6H_{12}	110-82-7	148, 218, 219	196, 220, 221	—
Ketone	Cyclohexanone	$C_6H_{10}O$	108-94-1	126, 214	174, 215	—
Cycloalkene	Cyclohexene	C_6H_{10}	110-83-8	137	186	—
Aryl versus Alkyl Series	Cyclohexene	C_6H_{10}	110-83-8	148	196	—
Cycloalkene	Cyclooctene	C_8H_{14}	931-87-3	137	186	—
Alkane	Cyclopentane	C_5H_{10}	287-92-3	132	—	—
Alkane & Cycloalkane Comparison	Cyclopentane	C_5H_{10}	287-92-3	132	180	—
Pentanol Series	Cyclopentanol	$C_5H_{10}O$	96-41-3	124	172	—
Cycloalkene	Cyclopentene	C_5H_8	142-29-0	137	186	—
Aldehyde	Decanal	$C_{10}H_{20}O$	112-31-2	125, 216	173, 217	—
Carboxylic Acid	Decanoic acid	$C_{10}H_{20}O_2$	334-48-5	—	231	—
Methyl Comparison Group	dichloromethyl Silane	CH_4Cl_2	75-54-7	135	183	—
Alkane	Dodecane	$C_{12}H_{26}$	112-40-3	130	178	—

Carboxylic Acid	Dodecanoic acid	$C_{12}H_{24}O_2$	143-07-7	154, 230	202	—
Amino Acid	D-Valine, 3-methyl, 1,1-dimethylethyl ester	$C_{10}H_{21}NO_2$	61169-85-5	146	194	—
Ether	ethoxy-Acetic acid	$C_4H_8O_3$	627-03-2	123	171	—
Ether	ethoxy-Ethene	C_4H_8O	109-92-2	123	171	—
Four-Carbon (Butyl) Comparison	Ethoxy-Ethene	C_4H_8O	109-92-2	127	—	—
Ether	ethoxy-Ethyne	C_4H_6O	927-80-0	123	171	—
Four-Carbon (Butyl) Comparison	Ethoxy-Ethyne	C_4H_6O	927-80-0	127	175	—
Polymers and Rubbers	Ethyl cellulose	polymer	N/A	—	—	161
Polymers and Rubbers	Ethylene Vinyl Acetate	polymer	N/A	—	—	159
Amide	Formamide	CH_3NO	75-12-7	144, 228	192, 229	—
Carboxylic Acid	Formic acid	CH_2O_2	64-18-6	154, 230	202, 231	—
Methyl Comparison Group	Formic acid, methyl ester	$C_2H_4O_2$	107-31-3	135	183	—
Cyclic Ether	Furan	C_4H_4O	110-00-9	152	200	—
Heterocyclic Compound	Furan	C_4H_4O	110-00-9	155	203	—
Aldehyde	Heptanal	$C_7H_{14}O$	111-71-7	125, 216	217	—
Alkane	Heptane	C_7H_{16}	142-82-5	130	178	—
Isomerism Comparison Group	Heptane	C_7H_{16}	142-82-5	134	182	—
Amide	hexamethyl-Phosphoric triamide	$C_6H_{18}N_3OP$	680-31-9	144	192	—
Aldehyde	Hexanal	$C_6H_{12}O$	66-25-1	—	173	—
Alkane	Hexane	C_6H_{14}	110-54-3	130	—	—
Aryl Amine	Indole, 1-H	C_8H_7N	120-72-9	143	191	—
Aromatic Compound with N or S	Indole, 1H-	C_8H_7N	120-72-9	147	195	—
Amino Acid	L-Valine, 3-methyl-, 1,1-dimethylethyl ester	$C_{10}H_{21}NO_2$	31556-74-8	146	194	—
Alcohol (mono-OH)	Methanol	CH_4O	67-56-1	119, 208, 209	167, 210	—
Methyl Comparison Group	Methanol	CH_4O	67-56-1	135	183	—
Ether	methoxy-Benzene	C_7H_8O	100-66-3	122, 212	170, 213	—
Carboxylic Acid	Methyl ester-Benzoic acid	$C_8H_8O_2$	93-58-3	154	202	—
Aryl versus Alkyl Series	methyl-Benzene	C_7H_8	108-88-3	149, 234, 235	166, 197, 236, 237	—
Aromatic (Aryl) Compound	methyl-Benzene	C_7H_8	108-88-3	118, 151	166, 199	—
Functional Group Comparisons	methyl-Benzene	C_7H_8	108-88-3	157	205	—
Alkane	Methylcyclopentane	C_6H_{12}	96-37-7	132	—	—

(continued)

Functional Group	Compound Name	Molecular Formula	CAS Number	10,500 – 6300 cm⁻¹ (952 – 1587 nm)	7200-3800 cm⁻¹ (1389 – 2632 nm)	Other NIR Regions
Alkane & Cycloalkane Comparison	methyl-Cyclopentane	C_6H_{12}	96-37-7	132	180	—
Amide	N, N-dimethyl-3-methyl-Benzamide	$C_{10}H_{13}NO$	—	145	193	—
Amide	N, N-dimethyl-3-oxo-Benzamide	$C_9H_{10}NO_2$	—	145	193	—
Alkyl Amine	N,N,N',N'-tetramethyl-1,4-Butanediamine	$C_8H_{20}N_2$	111-51-3	142	190	—
Amides, Comparison Group	N,N-diethyl-3-methyl-Benzamide	$C_{12}H_{17}NO$	134-62-3	—	193	—
Aryl Amine	N,N-dimethyl-1,3-Benzenediamine	$C_8H_{12}N_2$	2836-04-2	143	191	—
Amide	N,N-dimethyl-2-Propenamide	C_5H_9NO	2680-03-7	144	192	—
Amides, Comparison Group	N,N-dimethyl-3-oxo-Butanamide	$C_8H_{15}NO_2$	2235-46-3	—	193	—
Alkyl Amine	N,N-dipropyl-1-Propanamine	$C_9H_{21}N$	102-69-2	142	190	—
Aromatic (Aryl) Compound	Naphthalene	$C_{10}H_8$	91-20-3	151	199	—
Third Overtone C–H Comparison	n-Decane	$C_{10}H_{22}$	124-18-5	—	—	163, 238
Amino Acid	N-formyl-Glycine, ethyl ester	$C_5H_9NO_3$	4172-32-1	146	194	—
Aromatic (Aryl) Compound	nitro-Benzene	$C_6H_5NO_2$	98-95-3	151	199	—
Ethyl (C_2) Comparison Group	nitro-Ethane	$C_2H_5NO_2$	79-24-3	133	181	—
Methyl Comparison Group	nitro-Methane	CH_3NO_2	75-52-5	135	183	—
Amide	N-methyl Formamide	C_2H_5NO	123-39-7	145	193	—
Amide	N-methyl Propanamide	C_4H_9NO	187-58-2	145	193	—
Amide	N-methyl-Acetamide	C_3HH_7NO	79-16-3	144, 145, 228	192, 193, 229	—
Amides, Comparison Group	N-methyl-Acetamide	C_3H_7NO	79-16-3	153	193, 201	—
Acid vs. Amide vs. Ester Comparison	N-methyl-Acetamide	C_3H_7NO	79-16-3	153	—	—
Amides, Comparison Group	N-methyl-Formamide	C_2H_5NO	123-39-7	—	193	—
Amides, Comparison Group	N-methyl-Propanamide	C_4H_9NO	187-58-2	—	193	—
Alkane	Octadecane	$C_{18}H_{38}$	593-45-3	130	178	—
Alkane	Octane	C_8H_{18}	111-65-9	130	178	—
Aryl versus Alkyl Series	Octane	C_8H_{18}	111-65-9	148, 218, 219	196, 220, 221	—
Aldehyde	Pentanal	$C_5H_{10}O$	110-62-3	125	173	—
Alkane	Pentane	C_5H_{12}	109-66-0	130	178	—
Isomerism Comparison Group	Pentane	C_5H_{12}	109-66-0	134	182	—
X-H Compound Comparison	Pentane (C–C–H)	C_5H_{12}	109-66-0	141	189	—

Category	Compound name	Formula	CAS			
Carboxylic Acid	Pentanoic acid	$C_5H_{10}O_2$	109-52-4	—	202	—
Aromatic (Aryl) Compound	Phenanthrene	$C_{14}H_{10}$	85-01-8	151	199	—
Polymers and Rubbers	Poly(ethylene oxide)	polymer	N/A	—	—	161
Polymers and Rubbers	Poly(isobutyl methacrylate)	polymer	N/A	—	—	161
Polymers and Rubbers	Poly(vinyl butyral)	polymer	N/A	—	—	162
Polymers and Rubbers	Poly(vinyl chloride)	polymer	N/A	—	—	162
Polymers and Rubbers	Poly(vinyl pyrrolidone)	polymer	N/A	—	—	162
Polymers and Rubbers	Poly(vinyl stearate)	polymer	N/A	—	—	162
Polymers and Rubbers	Poly(vinylidene fluoride)	polymer	N/A	—	—	162
Polymers and Rubbers	Polyacrylic acid	polymer	N/A	—	—	160
Polymers and Rubbers	Polyethylene, chlorinated (25% Cl)	polymer	N/A	—	—	161
Polymers and Rubbers	Polypropylene, isotactic, chlorinated	polymer	N/A	—	—	161
Polymers and Rubbers	Polystyrene	polymer	N/A	—	—	160
Carboxylic Acid	Propanoic acid	$C_3H_6O_2$	79-09-4	154	—	—
Aromatic Compound with N or S	Quinoline	C_9H_7N	91-22-5	147	195	—
Heterocyclic Compound	Quinoline	C_9H_7N	91-22-5	155	203	—
Polymers and Rubbers	Silicone	polymer	N/A	—	—	160
Polymers and Rubbers	Starch	polymer	N/A	—	—	160
Polymers and Rubbers	Styrene isoprene styrene	polymer	N/A	—	—	160
Polymers and Rubbers	Styrene, ethylene, styrene copolymer	polymer	N/A	—	—	160
Third Overtone C–H Comparison	tert-Butanol	$C_4H_{10}O$	75-65-0	—	—	163, 238
Cyclic Ether	tetrahydro-2H-Pyran	$C_5H_{10}O$	142-68-7	152	200	—
Cyclic Ether	tetrahydro-4H-Pyran-4-one	$C_5H_8O_2$	29943-42-8	152	200	—
Aromatic Compound with N or S	Thiophene	C_4H_4S	110-02-1	147	195	—
Heterocyclic Compound	Thiophene	C_4H_4S	110-02-1	155	203	—
Third Overtone C–H Comparison	Toluene	C_7H_8	108-88-3	—	—	163, 238
Third Overtone C–H Comparison	Trimethyl pentane	C_8H_{18}	540-84-1	—	—	163, 238
Alkane	Undecane	$C_{11}H_{24}$	1120-21-4	118, 234, 235	166, 236, 237	—
Functional Group Comparisons	Undecane	$C_{11}H_{24}$	1120-21-4	157	205	—

Note: N/A indicates natural or synthetic product with no designated CAS number.

Appendix 7: Spectra Index
by Molecular Formula
(Carbon Number)

Functional Group	Compound Name	Molecular Formula	CAS Number	10,500 – 6300 cm⁻¹ (952 – 1587 nm)	7200-3800 cm⁻¹ (1389 – 2632 nm)	Other NIR Regions
Carboxylic Acid	Formic acid	CH_2O_2	64-18-6	154, 230	202, 231	—
Amide	Formamide	CH_3NO	75-12-7	144, 228	192, 229	—
Methyl Comparison Group	nitro-Methane	CH_3NO_2	75-52-5	135	183	—
Methyl Comparison Group	dichloromethyl Silane	CH_4Cl_2	75-54-7	135	183	—
Alcohol (mono-OH)	Methanol	CH_4O	67-56-1	119, 208, 209	167, 210	—
Methyl Comparison Group	Methanol	CH_4O	67-56-1	135	183	—
Ethyl (C₂) Comparison Group	1,2-dibromo-Ethane	$C_2H_4Br_2$	106-93-4	133	181	—
Ethyl (C₂) Comparison Group	1,1,1-trichloro-Ethane	$C_2H_4Cl_3$	71-55-6	133	181	—
Ethyl (C₂) Comparison Group	1,2-dichloro-Ethane	$C_2H_4Cl_2$	107-06-2	133	181	—
Methyl Comparison Group	Formic acid, methyl ester	$C_2H_4O_2$	107-31-3	135	183	—
Ethyl (C₂) Comparison Group	2-bromo-Ethanol	C_2H_5BrO	540-51-2	133	181	—
Amide	N-methyl Formamide	C_2H_5NO	123-39-7	145	193	—
Amides, Comparison Group	N-methyl-Formamide	C_2H_5NO	123-39-7	—	193	—
Ethyl (C₂) Comparison Group	nitro-Ethane	$C_2H_5NO_2$	79-24-3	133	181	—
Alcohol (diol)	1,2-Ethanediol	$C_2H_6O_2$	107-21-1	120	168	—
Alkyl Amine	1,2-Propanediamine	$C_3H_{10}N_2$	78-90-0	142, 224	190, 225	—
Aldehyde	2-Propynal	C_3H_2O	624-67-9	125	173	—
Alkyne	3-chloro-1-Propyne	C_3H_3Cl	624-65-7	140	188	—
Ketone	2-Propanone (Acetone)	C_3H_6O	67-64-1	126	174	—
Pentanol Series	2-Propene-1-ol	C_3H_6O	107-18-6	124	172	—
Third Overtone C–H Comparison	Acetone	C_3H_6O	67-64-1	—	—	163, 238
Carboxylic Acid	Propanoic acid	$C_3H_6O_2$	79-09-4	154	—	—
Propyl Comparison Group	1-chloro-Propane	C_3H_7Cl	540-54-5	136	184	—
Propyl Comparison Group	2-chloro Propane	C_3H_7Cl	75-29-6	136	184	—
Amides, Comparison Group	N-methyl-Acetamide	C_3H_7NO	79-16-3	153	193, 201	—
Acid vs. Amide vs. Ester Comparison	N-methyl-Acetamide	C_3H_7NO	79-16-3	153	—	—
Alcohol (mono-OH)	1-Propanol	C_3H_8O	71-23-8	119,209	—	—
Propyl Comparison Group	1-Propanol	C_3H_8O	71-23-8	136	184	—
Alcohol (mono-OH)	2-Propanol	C_3H_8O	67-63-0	119, 209	167	—
Propyl Comparison Group	2-Propanol	C_3H_8O	67-63-0	136	184	—
Alcohol (diol)	1,2 Propanediol	$C_3H_8O_2$	57-55-6	120	168	—
X-H Compound Comparison	2-Propanamine (C–NH)	C_3H_9N	75-31-0	141	189	—

Category	Compound	Formula	CAS			
Amide	N-methyl-Acetamide	C_3H_7NO	79-16-3	144, 145, 228	192, 193, 229	—
Alcohol (mono-OH)	1-Butanol	$C_4H_{10}O$	71-36-3	119, 208, 209	167, 210	—
Butyl Comparison Group	1-Butanol	$C_4H_{10}O$	71-36-3	131	179	—
Alcohol (mono-OH)	2-Butanol	$C_4H_{10}O$	78-92-2	209	—	—
Four-Carbon (Butyl) Comparison	2-Butanol	$C_4H_{10}O$	78-92-2	127	175	—
Butyl Comparison Group	2-Butanol	$C_4H_{10}O$	78-92-2	131	179	—
X-H Compound Comparison	2-Butanol (C-OH)	$C_4H_{10}O$	78-92-2	141	189	—
Alcohol (mono-OH)	2-methyl-2-Propanol	$C_4H_{10}O$	75-65-0	118, 119, 209, 234, 235	166,167, 236, 237	—
Functional Group Comparisons	2-methyl-2-Propanol	$C_4H_{10}O$	75-65-0	157, 234, 235	205, 236, 237	—
Third Overtone C-H Comparison	tert-Butanol	$C_4H_{10}O$	75-65-0	—	—	163, 238
Ether	1,1-dimethoxy-Ethane	$C_4H_{10}O_2$	534-15-6	122	170	—
Butanediol Series	1,2-Butanediol	$C_4H_{10}O_2$	584-03-2	121	169, 211	—
Butanediol Series	1,3-Butanediol	$C_4H_{10}O_2$	107-88-0	121	169, 211	—
Butanediol Series	1,4-Butanediol	$C_4H_{10}O_2$	110-63-4	121	169, 211	—
Alcohol (diol)	2,3-Butanediol	$C_4H_{10}O_2$	513-85-9	120, 210	168, 211	—
Butanediol Series	2,3-Butanediol	$C_4H_{10}O_2$	513-85-9	121, 208	169	—
Ether	2-ethoxy-Ethanol	$C_4H_{10}O_2$	110-80-5	123	171	—
Four-Carbon (Butyl) Comparison	2-ethoxy-Ethanol	$C_4H_{10}O_2$	110-80-5	127	175	—
X-H Compound Comparison	2-Butanethiol (C-SH)	$C_4H_{10}S$	513-53-1	141	189	—
X-H Compound Comparison	1-Butanamine (C-NH)	$C_4H_{11}N$	109-73-9	141	189	—
Cyclic Ether	Furan	C_4H_4O	110-00-9	152	200	—
Heterocyclic Compound	Furan	C_4H_4O	110-00-9	155	203	—
Acid vs. Amide vs. Ester Comparison	Acetic acid	$C_2H_4O_2$	64-19-7	153	201	—
Carboxylic Acid	Acetic acid	$C_2H_4O_2$	64-19-7	230	231	—
Aromatic Compound with N or S	Thiophene	C_4H_4S	110-02-1	147	195	—
Heterocyclic Compound	Thiophene	C_4H_4S	110-02-1	155	203	—
Alkyne	2-Butyne	C_4H_6	503-17-3	140	188	—
Polyfunctional Comparison	2-Butenal	C_4H_6O	123-73-9	128	176	—
Ether	ethoxy-Ethyne	C_4H_6O	927-80-0	123	171	—
Four-Carbon (Butyl) Comparison	Ethoxy-Ethyne	C_4H_6O	927-80-0	127	175	—
Heterocyclic (5-member ring)	4,5-dihydro-2(methylthio)-Thiazole	$C_4H_7NS_2$	19975-56-5	156	204	—
Ketone	2-Butanone	C_4H_8O	78-93-3	214	215	—
Four-Carbon (Butyl) Comparison	2-Butanone	C_4H_8O	78-93-3	127, 214	175	—
Aldehyde	Butanal	C_4H_8O	123-72-8	125, 216	173, 217	—
Ether	ethoxy-Ethene	C_4H_8O	109-92-2	123	171	—
Four-Carbon (Butyl) Comparison	Ethoxy-Ethene	C_4H_8O	109-92-2	127	171	—

(continued)

Functional Group	Compound Name	Molecular Formula	CAS Number	10,500 – 6300 cm⁻¹ (952 – 1587 nm)	7200-3800 cm⁻¹ (1389 – 2632 nm)	Other NIR Regions
Pentanol Series	2-Butene-1,4-diol	$C_4H_8O_2$	110-64-5	124	172	—
Four-Carbon (Butyl) Comparison	Butanoic Acid	$C_4H_8O_2$	107-92-6	127	175	—
Ether	ethoxy-Acetic acid	$C_4H_8O_3$	627-03-2	123	171	—
Butyl Comparison Group	2-bromo-Butane	C_4H_9Br	78-76-2	131	179	—
Butyl Comparison Group	1-chlorobutane	C_4H_9Cl	109-69-3	131	179	—
Butyl Comparison Group	2-chloro-Butane	C_4H_9Cl	78-86-4	131	179	—
Butyl Comparison Group	2-iodo-Butane	C_4H_9I	513-48-4	131	179	—
Amide	N-methyl Propanamide	C_4H_9NO	187-58-2	145	193	—
Amides, Comparison Group	N-methyl-Propanamide	C_4H_9NO	187-58-2	—	193	—
Alkene	1-Pentene	C_5H_{10}	109-67-1	138	185	—
Alkene, Internal versus Terminal	2-methyl-2-Butene	C_5H_{10}	513-35-9	139	187	—
Alkane	Cyclopentane	C_5H_{10}	287-92-3	132	—	—
Alkane & Cycloalkane Comparison	Cyclopentane	C_5H_{10}	287-92-3	132	180	—
Ketone	2-Pentanone	$C_5H_{10}O$	107-87-9	214	215	—
Pentanol Series	Cyclopentanol	$C_5H_{10}O$	96-41-3	124	172	—
Aldehyde	Pentanal	$C_5H_{10}O$	110-62-3	125	173	—
Cyclic Ether	tetrahydro-2H-Pyran	$C_5H_{10}O$	142-68-7	152	200	—
Carboxylic Acid	Pentanoic acid	$C_5H_{10}O_2$	109-52-4	—	202	—
Heterocyclic (5-member ring)	2,2-dimethyl-Thiazolidine	$C_5H_{11}NS$	19351-18-9	156	204	—
Alkane	Pentane	C_5H_{12}	109-66-0	130	178	—
Isomerism Comparison Group	Pentane	C_5H_{12}	109-66-0	134	182	—
X-H Compound Comparison	Pentane (C–C–H)	C_5H_{12}	109-66-0	141	189	—
Pentanol Series	1-Pentanol	$C_5H_{12}O$	71-41-0	124	172	—
Pentanol Series	2-Pentanol	$C_5H_{12}O$	6032-29-7	124	172	—
Alcohol (mono-OH)	3-Pentanol	$C_5H_{12}O$	584-02-01	119, 209	167	—
Pentanol Series	3-Pentanol	$C_5H_{12}O$	584-02-1	124	172	—
Ether	2,2-dimethoxy-Propane	$C_5H_{12}O_2$	77-76-7	122, 212	170, 213	—
Alcohol (diol)	2,4-Pentanediol	$C_5H_{12}O_2$	625-69-4	120	168	—
Ether	1,1,2-trimethoxy-Ethane	$C_5H_{12}O_3$	24332-20-5	122, 212	170, 213	—
Heterocyclic Compound	2-bromo-Pyridine	C_5H_4BrN	109-04-6	155	203	—
Polyfunctional Comparison	2-Furanmethanol	$C_5H_6O_2$	98-00-0	128	176	—
Aromatic Compound with N or S	1-methyl-1H-Pyrrole	C_5H_7N	96-54-8	147	195	—

Heterocyclic Compound	1-methyl-1H-Pyrrole	C_5H_7N	96-54-8	155	203	—
Alkyne	1-Pentyne	C_5H_8	627-19-0	140, 226	188, 227	—
Polyene	2-methyl-1,3-Butadiene	C_5H_8	78-79-5	158	206	—
Cycloalkene	Cyclopentene	C_5H_8	142-29-0	137	186	—
Cyclic Ether	3,4-dihydro-2H-Pyran	C_5H_8O	110-87-2	152	200	—
Polyfunctional Comparison	3-methyl-2-Butenal	C_5H_8O	107-86-8	128	176	—
Acid vs. Amide vs. Ester Comparison	1-Propen-2-ol acetate	$C_5H_8O_2$	108-22-5	153	201	—
Cyclic Ether	tetrahydro-4H-Pyran-4-one	$C_5H_8O_2$	29943-42-8	152	200	—
Amide	N,N-dimethyl-2-Propenamide	C_5H_9NO	2680-03-7	144	192	—
Amino Acid	N-formyl-Glycine, ethyl ester	$C_5H_9NO_3$	4172-32-1	146	194	—
Polyene	1,5-Hexadiene	C_6H_{10}	592-42-7	158, 232	206, 233	—
Cycloalkene	Cyclohexene	C_6H_{10}	110-83-8	137	186	—
Aryl versus Alkyl Series	Cyclohexene	C_6H_{10}	110-83-8	148	196	—
Ketone	Cyclohexanone	$C_6H_{10}O$	108-94-1	126, 214	174, 215	—
Alkene	1-Hexene	C_6H_{12}	592-41-6	138	185	—
Alkene, Internal versus Terminal	2,3-dimethyl-2-Butene	C_6H_{12}	563-79-1	139	187	—
Alkene, Internal versus Terminal	3,3-dimethyl-1-Butene	C_6H_{12}	558-37-2	139	187	—
Alkane	Cyclohexane	C_6H_{12}	110-82-7	132, 218	—	—
Alkane & Cycloalkane Comparison	Cyclohexane	C_6H_{12}	110-82-7	132	180	—
Aryl versus Alkyl Series	Cyclohexane	C_6H_{12}	110-82-7	148, 218, 219	196, 220, 221	—
Alkane	Methylcyclopentane	C_6H_{12}	96-37-7	132	—	—
Alkane & Cycloalkane Comparison	methyl-Cyclopentane	C_6H_{12}	96-37-7	132	180	—
Ketone	2-Hexanone	$C_6H_{12}O$	591-78-6	126	174	—
Ketone	3-Hexanone	$C_6H_{12}O$	589-38-8	118, 126	166, 174	—
Functional Group Comparisons	3-Hexanone	$C_6H_{12}O$	589-38-8	157, 234, 235	205, 236, 237	—
Aldehyde	Hexanal	$C_6H_{12}O$	66-25-1	—	173	—
Acid vs. Amide vs. Ester Comparison	2-ethoxy- acetate-Ethanol	$C_6H_{12}O_3$	111-15-9	153	201	—
Isomerism Comparison Group	2-methylpentane	C_6H_{14}	107-83-5	134	182	—
Alkane	Hexane	C_6H_{14}	110-54-3	130	—	—
Ether	1,2-diethoxy-Ethane	$C_6H_{14}O_2$	629-14-1	122	170	—
Alcohol (diol)	2,5-Hexanediol	$C_6H_{14}O_2$	2935-44-6	120	168	—
Amide	hexamethyl-Phosphoric triamide	$C_6H_{18}N_3OP$	680-31-9	144	192	—
Aromatic (Aryl) Compound	2,4,5-trichloro-Phenol	$C_6H_3Cl_3O$	95-95-4	150	198	—
Aromatic (Aryl) Compound	bromo-Benzene	C_6H_5Br	108-86-1	151	199	—

(continued)

Functional Group	Compound Name	Molecular Formula	CAS Number	10,500 – 6300 cm⁻¹ (952 – 1587 nm)	7200-3800 cm⁻¹ (1389 – 2632 nm)	Other NIR Regions
Aromatic (Aryl) Compound	nitro-Benzene	$C_6H_5NO_2$	98-95-3	151	199	—
Aryl versus Alkyl Series	Benzene	C_6H_6	71-43-2	149, 218, 219	197, 220, 221	—
Functional Group Comparisons	Benzene	C_6H_6	71-43-2	157	205	—
Aryl Amine	4-methyl-Pyridine	C_6H_7N	108-89-4	143	191	—
Aromatic Compound with N or S	4-methyl-Pyridine	C_6H_7N	108-89-4	147	195	—
Aryl Amine	Benzenamine	C_6H_7N	62-53-3	143	191	—
Aryl Amine	2-Pyridinemethanol	C_6H_7NO	586-98-1	143	191	—
Polyene	1,3-Cyclohexadiene	C_6H_8	592-57-4	158	206	—
Ketone	2-Cyclohexen-1-one	C_6H_8O	930-68-7	126	174	—
Heterocyclic (5-member ring)	4,5-dihydro-2-(2-propenylthio)-Thiazole	$C_6H_9NS_2$	3571-74-2	156	204	—
Alkyne	1-Heptyne	C_7H_{12}	628-71-7	226	227	—
Cycloalkene	Cycloheptene	C_7H_{12}	628-92-2	137	186	—
Alkene	1-Heptene	C_7H_{14}	592-76-7	222	223	—
Alkane	Cycloheptane	C_7H_{14}	291-64-5	132	—	—
Alkane & Cycloalkane Comparison	Cycloheptane	C_7H_{14}	291-64-5	132	180	—
Aldehyde	Heptanal	$C_7H_{14}O$	111-71-7	125, 216	217	—
Alkane	Heptane	C_7H_{16}	142-82-5	130	178	—
Isomerism Comparison Group	Heptane	C_7H_{16}	142-82-5	134	182	—
Alkyl Amine	1,7-Heptanediamine	$C_7H_{18}N_2$	646-19-5	142	190	—
Aldehyde, Substituted Aromatic	2-bromo-Benzaldehyde	C_7H_5BrO	6630-33-7	129	177	—
Aldehyde, Substituted Aromatic	2-chloro-Benzaldehyde	C_7H_5ClO	89-98-5	129	177	—
Aldehyde, Substituted Aromatic	Benzaldehyde	C_7H_6O	100-52-7	129	177	—
Aryl versus Alkyl Series	Benzaldehyde	C_7H_6O	100-52-7	149	197	—
Polyene	1,3,5-Cycloheptatriene	C_7H_8	544-25-2	158	206	—
Aryl versus Alkyl Series	methyl-Benzene	C_7H_8	108-88-3	149, 234, 235	166, 197, 236, 237	—
Aromatic (Aryl) Compound	methyl-Benzene	C_7H_8	108-88-3	118, 151	166, 199	—
Functional Group Comparisons	methyl-Benzene	C_7H_8	108-88-3	157	205	—
Third Overtone C–H Comparison	Toluene	C_7H_8	108-88-3	—	—	163, 238
Aromatic (Aryl) Compound	3-methyl-Phenol	C_7H_8O	08-39-4	150	198	—
Aromatic (Aryl) Compound	4-methyl-Phenol	C_7H_8O	106-44-5	151	199	—
Ether	methoxy-Benzene	C_7H_8O	100-66-3	122, 212	170, 213	—

Category	Compound	Formula	CAS Number				
Heterocyclic Compound	3-ethyl-Pyridine	C₇H₉N	536-78-7	155		203	—
Aromatic (Aryl) Compound	1,2-dimethyl-Benzene (xylene)	C₈H₁₀	95-47-6	150	—	198	—
Aromatic (Aryl) Compound	1,3-dimethyl-Benzene (xylene)	C₈H₁₀	108-38-3	150		198	—
Aromatic (Aryl) Compound	1,4-dimethyl-Benzene (xylene)	C₈H₁₀	106-42-3	150		198	—
Aromatic (Aryl) Compound	2-ethyl-Phenol	C₈H₁₀O	90-00-6	150		171	—
Ether	2-phenoxy-Ethanol	C₈H₁₀O₂	122-99-6	123		206	—
Polyene	1,3-Cyclooctadiene	C₈H₁₂	1700-10-3	158		191	—
Aryl Amine	N,N-dimethyl-1,3-Benzenediamine	C₈H₁₂N₂	2836-04-2	143		196	—
Aryl versus Alkyl Series	1,7-Octadiene	C₈H₁₄	3710-30-3	148		233	—
Polyene	1,7-Octadiene	C₈H₁₄	3710-30-3	232		188, 227	—
Alkyne	1-Octyne	C₈H₁₄	629-05-0	140, 226		188	—
Alkyne	4-Octyne	C₈H₁₄	1942-45-6	140		186	—
Cycloalkene	Cyclooctene	C₈H₁₄	931-87-3	137		193	—
Amides, Comparison Group	N,N-dimethyl-3-oxo-Butanamide	C₈H₁₅NO₂	2235-46-3		—		—
Alkene	1-Octene	C₈H₁₆	111-66-0	138		185	—
Isomerism Comparison Group	2,2,4-trimethylpentane	C₈H₁₈	540-84-1	134		182	—
Aryl versus Alkyl Series	2,2,4-trimethyl-Pentane	C₈H₁₈	540-84-1	118, 148, 218, 219, 234, 235		166, 196, 220, 221, 236, 237	—
Functional Group Comparisons	2,2,4-trimethyl-Pentane	C₈H₁₈	540-84-1	157		205	—
Alkane	Octane	C₈H₁₈	111-65-9	130		178	—
Aryl versus Alkyl Series	Octane	C₈H₁₈	111-65-9	148, 218, 219		196, 220, 221	—
Third Overtone C–H Comparison	Trimethyl pentane	C₈H₁₈	540-84-1		—		163, 238
Alkyl Amine	N,N,N',N'-tetramethyl-1,4-Butanediamine	C₈H₂₀N₂	111-51-3	142		190	—
Heterocyclic Compound	1H-Indole	C₈H₇N	120-72-9	155		203	—
Aryl Amine	Indole, 1-H	C₈H₇N	120-72-9	143		191	—
Aromatic Compound with N or S	Indole, 1H-	C₈H₇N	120-72-9	147		195	—
Carboxylic Acid	Methyl ester-Benzoic acid	C₈H₈O₂	93-58-3	154		202	—
Amide	N, N-dimethyl-3-oxo-Benzamide	C₉H₁₀NO₂		145		193	—
Aldehyde, Substituted Aromatic	2,4-dimethyl-Benzaldehyde	C₉H₁₀O	15764-16-6	129		177	—
Polyfunctional Comparison	Benzenepropanal	C₉H₁₀O	104-53-0	128		176	—
Alkene	1-Nonene	C₉H₁₈	124-11-8	222		223	—
Alkyl Amine	N,N-dipropyl-1-Propanamine	C₉H₂₁N	102-69-2	142		190	—
Aromatic Compound with N or S	Quinoline	C₉H₇N	91-22-5	147		195	—
Heterocyclic Compound	Quinoline	C₉H₇N	91-22-5	155		203	—

(continued)

Functional Group	Compound Name	Molecular Formula	CAS Number	10,500 – 6300 cm⁻¹ (952 – 1587 nm)	7200-3800 cm⁻¹ (1389 – 2632 nm)	Other NIR Regions
Polyfunctional Comparison	3-phenyl-2-Propenal	C_9H_8O	14371-10-9	128	176	—
Aldehyde, Substituted Aromatic	2,4,6-trimethyl-Benzaldehyde	$C_{10}H_{12}O$	487-68-3	129	177	—
Amide	N, N-dimethyl-3-methyl-Benzamide	$C_{10}H_{13}NO$	—	145	193	—
Alkene	1-Decene	$C_{10}H_{20}$	872-05-9	138, 222	185, 223	—
Aldehyde	Decanal	$C_{10}H_{20}O$	112-31-2	125, 216	173, 217	—
Carboxylic Acid	Decanoic acid	$C_{10}H_{20}O_2$	334-48-5	—	231	—
Amino Acid	D-Valine, 3-methyl, 1,1-dimethylethyl ester	$C_{10}H_{21}NO_2$	61169-85-5	146	194	—
Amino Acid	L-Valine, 3-methyl-, 1,1- dimethylethyl ester	$C_{10}H_{21}NO_2$	31556-74-8	146	194	—
Third Overtone C–H Comparison	n-Decane	$C_{10}H_{22}$	124-18-5	—	—	163, 238
Alkyl Amine	1-Decanamine	$C_{10}H_{23}N$	2016-57-1	118, 142, 224, 234, 235	166, 190, 225, 236, 237	—
Functional Group Comparisons	1-Decanamine	$C_{10}H_{23}N$	2016-57-1	157, 234	205	—
Aromatic (Aryl) Compound	Naphthalene	$C_{10}H_8$	91-20-3	151	199	—
Alkane	Undecane	$C_{11}H_{24}$	1120-21-4	118, 234, 235	166, 236, 237	—
Functional Group Comparisons	Undecane	$C_{11}H_{24}$	1120-21-4	157	205	—
Aromatic (Aryl) Compound	1,1-Biphenyl	$C_{12}H_{10}$	92-52-4	150	198	—
Amides, Comparison Group	N,N-diethyl-3-methyl-Benzamide	$C_{12}H_{17}NO$	134-62-3	—	193	—
Carboxylic Acid	Dodecanoic acid	$C_{12}H_{24}O_2$	143-07-7	154, 230	202	—
Alkyl Amine	Cyclododecanamine	$C_{12}H_{25}N$	1502-03-0	142, 224	190, 225	—
Alkane	Dodecane	$C_{12}H_{26}$	112-40-3	130	178	—
Aromatic (Aryl) Compound	Phenanthrene	$C_{14}H_{10}$	85-01-8	151	199	—
Carboxylic Acid	Benzoic acid anhydride	$C_{14}H_{10}O_3$	93-97-0	230	231	—
Alkene	1-Tetradecene	$C_{14}H_{28}$	1120-36-1	138	185	—
Alkane	Octadecane	$C_{18}H_{38}$	593-45-3	130	178	—
Polymers and Rubbers	60% polypropylene and 40% polyester	polymer	N/A	—	—	159
Polymers and Rubbers	60% polypropylene-polyethylene acopolymer and 40% polyester	polymer	N/A	—	—	159
Polymers and Rubbers	Atactic Polypropylene	polymer	N/A	—	—	159
Polymers and Rubbers	Cellulose acetate	polymer	N/A	—	—	161

Polymers and Rubbers	Cellulose acetate butyrate	polymer	N/A	—	161
Polymers and Rubbers	Cellulose propionate	polymer	N/A	—	162
Polymers and Rubbers	Ethyl cellulose	polymer	N/A	—	161
Polymers and Rubbers	Ethylene Vinyl Acetate	polymer	N/A	—	159
Polymers and Rubbers	Poly(ethylene oxide)	polymer	N/A	—	161
Polymers and Rubbers	Poly(isobutyl methacrylate)	polymer	N/A	—	161
Polymers and Rubbers	Poly(vinyl butyral)	polymer	N/A	—	162
Polymers and Rubbers	Poly(vinyl chloride)	polymer	N/A	—	162
Polymers and Rubbers	Poly(vinyl pyrrolidone)	polymer	N/A	—	162
Polymers and Rubbers	Poly(vinyl stearate)	polymer	N/A	—	162
Polymers and Rubbers	Poly(vinylidene fluoride)	polymer	N/A	—	162
Polymers and Rubbers	Polyacrylic acid	polymer	N/A	—	160
Polymers and Rubbers	Polyethylene, chlorinated (25% Cl)	polymer	N/A	—	161
Polymers and Rubbers	Polypropylene, isotactic, chlorinated	polymer	N/A	—	161
Polymers and Rubbers	Polystyrene	polymer	N/A	—	160
Polymers and Rubbers	Silicone	polymer	N/A	—	160
Polymers and Rubbers	Starch	polymer	N/A	—	160
Polymers and Rubbers	Styrene isoprene styrene	polymer	N/A	—	160
Polymers and Rubbers	Styrene, ethylene, styrene copolymer	polymer	N/A	—	160

Note: N/A indicates natural or synthetic product with no designated CAS number

Index